IMPROVED HYDROLOGICAL UNDERSTANDING OF A SEMI-ARID SUBTROPICAL TRANSBOUNDARY BASIN USING MULTIPLE TECHNIQUES – THE INCOMATI RIVER BASIN

Aline Maraci Lopes Saraiva Okello

Cover page: View of the Sabie River from the Kruger National Park. Photo by Frank Eckardt.

IMPROVED HYDROLOGICAL UNDERSTANDING OF A SEMI-ARID SUBTROPICAL TRANSBOUNDARY BASIN USING MULTIPLE TECHNIQUES – THE INCOMATI RIVER BASIN

DISSERTATION

Submitted in fulfillment of the requirements of
the Board for Doctorates of Delft University of Technology
and
of the Academic Board of the IHE Delft
Institute for Water Education
for
the Degree of DOCTOR
to be defended in public on
Thursday, 2 May 2019, 15:00 hours
in Delft, the Netherlands

by

Aline Maraci Lopes SARAIVA OKELLO

Master of Science in Hydrology and Water Resources,
IHE Delft Institute for Water Education, Delft, The Netherlands
Born in Maputo, Mozambique

This dissertation has been approved by the promotors and copromotor

Composition of the Doctoral Committee:

Rector Magnificus TU Delft	Chairman
Rector IHE Delft	Vice-Chairman
Prof. dr. S. Uhlenbrook	IHE Delft / Delft University of Technology, promotor
Prof. dr. ir. P. van der Zaag	IHE Delft / Delft University of Technology, promotor
Dr. I. Masih	IHE Delft, copromotor

Independent members:

Prof. dr. D. Mazvimavi	University of Western Cape, South Africa
Dr. ir. H. C. Winsemius	Delft University of Technology
Prof. dr. ir. A. B. K. van Griensven	Vrije Universiteit Brussels, Belgium / IHE Delft
Prof. dr. M. E. McClain	IHE Delft and Delft University of Technology
Prof. dr. ir. N.C. van de Giesen	Delft University of Technology, reserve member

Prof. dr. G.P.W. Jewitt (University of KwaZulu-Natal, South Africa) significantly contributed towards the supervision of this dissertation.

This research was conducted under the auspices of the Graduate School for Socio-Economic and Natural Sciences of the Environment (SENSE)

CRC Press/Balkema is an imprint of the Taylor & Francis Group, an informa business

Published by:
CRC Press/Balkema
Schipholweg 107C, 2316 XC, Leiden, the Netherlands
Pub.NL@taylorandfrancis.com
www.crcpress.com – www.taylorandfrancis.com
ISBN: 978-0-367-28075-8

Acknowledgments

This Ph.D. project was supported by a Ph.D. fellowship grant from Netherlands Ministry of Development Cooperation (DGIS) through the UNESCO-IHE Partnership Research Fund (UPaRF). The study was carried out in the framework of the Research Project "Risk-based operational water management for the Incomati River Basin" (RISKOMAN) project. The Water Resources Commission (WRC) of South Africa, also contributed additional funding, through RISKOMAN-WRC part of the project K5/1935, managed by the University of KwaZulu Natal (UKZN, South Africa). All partners of the RISKOMAN project, IHE Delft (Netherlands), UKZN, Inkomati-Usuthu Catchment Management Agency (IUCMA, formerly ICMA, South Africa), Komati Basin Water Authority (KOBWA, Swaziland), Eduardo Mondlane University (UEM, Mozambique), including the reference group are thanked for their valuable inputs.

The financial contribution through research grants from the International Foundation for Science (IFS) Stockholm, Sweden, through a grant (W/5340-1), the L'Oréal-UNESCO "For Women in Science, Sub-Saharan Africa 2013" and Schlumberger Foundation "Faculty of the Future Fellowship 2014-2015" are also highly appreciated.

I would like to thank my promoters Prof. Dr. Stefan Uhlenbrook, Prof. Dr. Pieter van der Zaag, Prof. Dr. Graham Jewitt and supervisor Dr. Ilyas Masih for the patience, enthusiastic supervision, guidance, inspiration and encouragement to pursue this PhD. I also remember Prof. Huub Savenije, Prof. Diniz Juizo and Prof. Wim Bastiaanssen (WATPLAN project) for the valuable advice and contributions. It is such a privilege to work with extraordinary scholars like you.

The cooperation of the Soil Science Department of UKZN, particularly of Prof. Simon Lorentz and Mr. Cobus has been essential for the success of fieldwork and is gratefully acknowledged. The South African National Parks (SANParks) - Kruger National Park (KNP) is also recognized for allowing fieldwork to take place within the park area. To the staff of KNP, IUCMA and Department of Water and Sanitation (DWS) that helped with fieldwork my special thank you – it was a great and enriching experience for me to be out in the field with you. A special thanks to late

Maduna and the researchers of SANParks – KNP that have assisted with monthly collection of water samples. IHE-Delft laboratory staff is also thanked for assistance with analysis of isotopes, cations and anions.

Streamflow, climate and water quality data was kindly provided by the Department of Water and Sanitation (DWS) in South Africa, South African Weather Service (SAWS) and UKZN. Staff from Ara-Sul DGUBI is also acknowledged for data provided for Mozambican locations, as well as assistance with water sampling.

I would like to acknowledge the support and guidance of my colleagues at UKZN; Eddie Riddell, Tinisha, Mercy and many other students for their suggestions, discussions, and support during fieldwork. I appreciate a lot the support from various staff members of UKZN in gathering data and taking care of logistics. My appreciation also goes to the PhD colleagues at IHE Delft (former UNESCO-IHE) for the valuable ideas. The constructive comments of anonymous reviewers and editors from journals where several chapters of this PhD were published are greatly appreciated.

My heartfelt appreciation to several friends that helped me in various ways. Micah for countless advice regarding coding and modelling as well as Caroline, Veronica, Patricia, Chris, Pedi, Raquel, Vivian, and all the students that worked on Incomati/Riskoman project. The various staff members of IHE-Delft that always ensured we had pleasant and productive times in Delft. And to the Mozambican and Kenyan communities in Delft, who were my home away from home in the Netherlands.

Cloverley, my accountability 'buddy' for the last 3 years, really helped me to keep motivated and enthusiastic about my work; a big thank you also to Olga, from Productivity for Scientists.

The ladies Bible Study group in Richards Bay (South Africa), especially Wendy and Sue, have prayed countless times regarding my PhD and related issues. I really appreciate your friendship and fellowship. The connect group in Richards Bay, Richards Bay Community Church (RBCC) and Mount Zion International Parish (MZIP, Delft) communities, pastors, and prayer groups are also thanked for their support through the years.

Finally yet importantly, I would like to thank my dearest husband Nick for all support, encouragement, and patience. I could have never made it this far without his unconditional love and affection, protection and provision. To my daughters Zawadi, Priscilla and Nicole for surviving many days without mommy close by, and for going in several trips because of mommy's PhD. To my dear mother for the amazing support to my family while I was away, my siblings Melanie and Marcel, and for my entire family and friends in Mozambique and Kenya. All glory and honour to God!

To my family and all the brave women in this world!

Summary

Water scarcity is a major issue in today's agenda. For sub-Saharan Africa in particular, dramatic water shortages are predicted in the coming years. Effective water management is crucial for better use of available water resources. Understanding the hydrology of a catchment is important in order to support optimal use of water resources in the face of socio-economic and environmental constraints and uncertainties.

In heavily committed river basins such as the Incomati, with many different water use(r)s located in different riparian countries (South Africa, Swaziland and Mozambique), strong interdependencies exist. Water allocation decisions thus have important economic, social, environmental and political consequences. In this case, water allocation decision-making involves difficult trade-offs, for which decision-support tools exist that are frequently based on optimising an economic objective function subject to constraints representing, among other things, hydrological processes. Large variability of rainfall, both within and between years, leads to even larger variations in river flow and adds uncertainty to the water allocation equation.

This research project aims at improving understanding of hydrological processes of a river basin, particularly runoff generation processes, to enable better water management. Several tools and approaches were used to achieve this aim. Comprehensive statistical and trend analyses of rainfall and streamflow were conducted, and the Indicators of Hydrological Alteration (IHA) tool was used to describe the streamflow regime and trends over time (over 50 years of data were analysed). Significant trends in streamflow were mapped and correlated to potential drivers. Land use and land cover change, particularly the conversion of natural vegetation into forest plantation, the expansion of irrigated agriculture and flow regulation due to dam operation were identified as critical drivers of flow regime alteration in the Incomati basin.

Intensive fieldwork campaigns using tracer methods, particularly environmental isotopes were employed to improve understanding of runoff generation processes. A snapshot sampling of Incomati River system was conducted during wet and dry seasons of 2011 to 2013. In the wet season of 2013/2014, intense event sampling took place in a selected sub-catchment of the Incomati, the Kaap catchment. The fieldwork yielded understanding of major patterns of water quality in the basin, and their relationship with hydrological processes. A new data set of isotope and hydrochemistry data was generated, and can be used as baseline for further analysis in the basin, or for other studies of semi-arid subtropical catchments.

Hydrograph separation using long-term hydrochemical data at seasonal scale, and hydrochemical and isotope data at event scale were performed to quantify runoff components in the Kaap catchment. Sources of runoff and temporal dynamics of runoff generation were also described. Furthermore, a novel methodology to calibrate recursive digital filters using routinely collected water quality data was tested in the catchment. This method allows for estimation of daily baseflow components from daily streamflow data, which is not available in the catchment. This information is important for the operational water management in the catchment.

Finally, dominant runoff generation zones were mapped using the novel Height Above Nearest Drainage (HAND) approach, combined with knowledge of the geology of the Kaap catchment. The hydrological model STREAM was then employed, informed by the runoff generation zones mapping and the process understanding gathered in the catchment. Data to drive the model was carefully selected from the best available datasets (ground and remote sensing data) and local knowledge of the region. Model results highlighted further gaps in knowledge of hydrological processes in the catchment, and the need to improve the simulation of water abstractions and evaporation processes in the catchment.

The key results and insights of this research include:

- Changes in flow regime in the Incomati basin are mostly driven by anthropogenic activities (e.g. irrigated agriculture, forestation, dam operation) and not by climate change. This means that great attention should be put into land use planning and management, and overall water management in the basin, to ensure sustainable use of water resources, whilst protecting the environment.
- Anthropogenic activities also affect negatively water quality in the basin. While some stakeholders are already implementing measures to control water pollution, more emphasis is required to monitor and control pollution from various point and non-point sources. For this, it is critical that frequency of water quality monitoring is increased, and some real time water quality sensors are installed at key/hot spot locations.

- There is a good monitoring network for water quality in South Africa, however, there is an alarming trend of decrease in frequency of water sampling (from weekly to once in a month, or quarterly). Mozambique and Swaziland in contrast need to improve their monitoring networks, and could benefit from cooperation with South Africa.
- Routinely collected water quality data can be used to calibrate recursive digital filters, for separation of baseflow at daily time steps from daily streamflow data. This information is useful for operational quantification of environmental flows, management of water resources, and to inform and improve hydrological models in the region.
- Tracer investigations can be very costly, but with adequate project design they provide great insights into the dynamics of runoff generation in catchments.
- A strong correlation between antecedent precipitation index and direct runoff was found for several events in the Kaap catchment. This is likely the case for other semi-arid subtropical catchments. Wet conditions prior to rainfall events fill up storages in the catchment and result in higher contributions of direct runoff. This results in quicker response of the catchment to rainfall events.
- Landscape mapping, using approaches such as HAND, is useful to extract more information of available data in the catchment. Furthermore, there are several new/recent data sources that can greatly assist in modelling transboundary basins, such as the Soil grids 250m soil data base, remote sensing precipitation data (CHIRPS), and remote sensing evaporation estimates (ALEXI, CMERST, SSEBop, etc). Downscaling of these products in space and time as well as bias correction should be pursued as avenues to improve model inputs.
- The use of Soil grids 250m dataset yielded hydrological modelling results as good as local soil data – this is a promising avenue for the modelling of the Incomati and other data poor catchments, given that this is a freely available dataset.

The main recommendations of this study are:

- The strengthening of monitoring networks of rainfall, streamflow, groundwater and water quality, especially in Mozambique and Swaziland, is necessary. It is recommended that the countries share databases and follow similar protocols for data collection and reporting.

- To conduct a careful basin-wide assessment of all benefits derived from water use, and re-assess the first priority water uses in the Incomati basin.
- To perform validation studies in semi-arid catchments to assess if regionalization (transfer in space) of recursive digital filter parameters is possible, using high frequency water quality data. This could be achieved with the installation of real time EC sensors in selected sub catchments for detailed calibration/validation exercises.
- To undertake a comprehensive study on bias correction, downscaling and calibration of remote sensing precipitation and actual evaporation data to use as input for hydrological modelling in semi-arid basins. This is particularly relevant in transboundary basins such as the Incomati, with uneven datasets.

The most important scientific innovation of this thesis is the application of water quality data to quantify and improve the understanding of runoff components in a semi-arid subtropical catchment. Furthermore, the testing of a method to calibrate recursive digital filters using readily available water quality data is an important step in improving the quantification of baseflow components, used to define environmental flows. Furthermore, several parameters were identified for hydrograph separation in semi-arid environments, which contribute to the scientific knowledge for such systems. Parameters for the STREAM model applied to the Kaap catchment also serve as benchmark for other similar catchments, as well as the process followed to improve model simulations, by using process studies, landscape mapping and improved data sources.

The relevance of this study to society is that with increasing pressure in water resources, this thesis presents a comprehensive assessment of water resources availability and variability in the Incomati River basin. The recently concluded Progressive Realisation of the Inco-Maputo Agreement (PRIMA) project proposed several IWRM strategies and plans, to address the challenges of water resources management in the Incomati. The riparian countries plan to increase storage capacity by building new dams and enlarging existing ones. Several water resources development projects are also planned, including the increase of irrigated agriculture and commercial forestry, and the abstraction of water for municipalities and to augment the city of Maputo water supply. However, the knowledge of the temporal variability of water resources and particularly the contribution of groundwater is not well understood. This issue is important in the implementation of near-real time management of environmental flows. Furthermore, the datasets used and produced in previous studies still lacked comprehensive understanding of runoff generation processes in the basin. This research addresses these gaps, by using multiple methods to better understand hydrological processes in the basin.

This research helps shedding some light on hotspots were current land use changes are impacting water resources availability. The research reviewed several studies conducted in the region, and provides a great starting point for conducting research in the Incomati River basin. The testing of new methods to make best use of available data is of high importance in this region, because water managers have to work with limited datasets, and being able to extract the most value out of routinely collected data is a great added value. In particular, the quantification of runoff components, through hydrograph separation, can be useful for environmental flows determination and low flow management.

Samenvatting

Waterschaarste is een belangrijk onderwerp op de wereld agenda. Vooral voor Afrika bezuiden de Sahara wordt de komende jaren een dramatisch watertekort voorspeld. Effectief waterbeheer is cruciaal voor een beter gebruik van de beschikbare waterbronnen. Het begrijpen van de hydrologie van een stroomgebied is belangrijk om het optimaal gebruik van water te ondersteunen in het licht van sociaaleconomische en ecologische belangen en onzekerheden.

In rivieren met een grote vraag naar water zoals de Incomati rivier in Zuidelijk Afrika, met veel en diverse watergebruikers in verschillende oeverstaten (Zuid-Afrika, Swaziland en Mozambique), bestaan er sterke onderlinge afhankelijkheden. Beslissingen hoe het schaarse water te verdelen kunnen dan belangrijke economische, sociale, ecologische en politieke gevolgen hebben. In de besluitvorming over watertoewijzing gaat het om complexe afwegingen, waarvoor beslissingsondersteunende instrumenten bestaan die vaak gebaseerd zijn op het optimaliseren van een economische doelstelling, welke onderhevig is aan bepaalde beperkingen, inclusief hydrologische processen. Grote variaties in regenval, zowel binnen als tussen jaren, leiden tot nog grotere variaties in rivierafvoeren en vergroten de onzekerheid in het waterverdelingsvraagstuk.

Dit onderzoeksproject is gericht op een beter begrip van de hydrologische processen van het stroomgebied, in het bijzonder afvoerprocessen, om beter waterbeheer mogelijk te maken. Verschillende instrumenten en benaderingen werden gebruikt om dit doel te bereiken. Er werden uitgebreide statistische en trendanalyses van regenval en rivierafvoer uitgevoerd en het *Indicators of Hydrological Alteration (IHA)* instrument werd ingezet om het afvoerregime en trends in de tijd te beschrijven (een periode langer 50 jaar werd geanalyseerd). Significante trends in rivierafvoer zijn in kaart gebracht en gecorreleerd aan potentiële oorzaken. Landgebruik en veranderingen in bodembedekking, met name de conversie van natuurlijke vegetatie naar houtplantages, de uitbreiding van geïrrigeerde landbouw en de stroomregulering als gevolg van stuwdammen, werden geïdentificeerd als belangrijkste oorzaken van de verandering van het afvoerregime van de Incomati-rivier.

Gedurende intensieve veldcampagnes werden zogenaamde tracer methoden, in het bijzonder omgevingsisotopen, gebruikt om het begrip van afvoerprocessen te verbeteren. Momentopnames van het Incomati-rivier systeem werden uitgevoerd tijdens de natte en droge seizoenen van 2011 tot 2013. In het natte seizoen van 2013/2014 vond een intense gebeurtenisbemonstering plaats in een geselecteerd deelstroomgebied van de Incomati, namelijk de Kaap. Het veldwerk gaf inzicht in belangrijke patronen van waterkwaliteit in het stroomgebied en hun relatie met hydrologische processen. Een nieuwe dataset van isotopen en hydrochemische gegevens werd gegenereerd en kan worden gebruikt als basis voor verdere analyse van het stroomgebied, of voor andere studies van semi-aride subtropische stroomgebieden.

Hydrograaf-separatie met behulp van hydrochemische gegevens op seizoensschaal, en hydrochemische en isotoopgegevens op gebeurtenisschaal, maakte het mogelijk de afvoercomponenten van de Kaap rivier te kwantificeren. De oorsprong van afvoer en de temporele dynamiek van afvoergeneratie werden ook beschreven. Verder werd in het stroomgebied een nieuwe methodologie getest voor het kalibreren van recursieve digitale filters met behulp van routinematig verzamelde waterkwaliteit gegevens. Met deze methode kunnen componenten van de lage rivier afvoer gedurende de droge tijd, de *baseflow,* worden geschat op basis van dagelijkse afvoergegevens, die tot dusver niet beschikbaar waren in het stroomgebied. Deze informatie is belangrijk voor het operationele waterbeheer in het stroomgebied.

Ten slotte werden dominante afvoergeneratie-zones in kaart gebracht met behulp van de nieuwe *Height Above Nearest Drainage (HAND)* benadering, gecombineerd met kennis van de geologie van het stroomgebied van de Kaap. Het hydrologische model STREAM werd vervolgens gebruikt en voorzien met de afvoergeneratie-zones en de procesbegrip verkregen in het stroomgebied. Gegevens voor het model werden zorgvuldig geselecteerd uit de best beschikbare datasets (grond- en satelliet-data) en lokale kennis van de regio. Modelresultaten identificeerden hiaten in de kennis van hydrologische processen in het stroomgebied en de noodzaak om de simulatie van wateronttrekkingen en verdampingsprocessen in het stroomgebied te verbeteren.

De belangrijkste resultaten en inzichten van dit onderzoek zijn:

- Veranderingen in het afvoerregime in de Incomati rivier worden meestal bepaald door antropogene activiteiten (bijvoorbeeld geïrrigeerde landbouw, houtplantages, stuwdammen) en niet door klimaatverandering. Dit betekent dat er veel aandacht moet worden besteed aan ruimtelijke ordening en -beheer, en aan het algehele waterbeheer in het stroomgebied, om een duurzaam gebruik van waterbronnen te waarborgen en tegelijkertijd het milieu te beschermen.
- Antropogene activiteiten hebben ook een negatief effect op de waterkwaliteit in het stroomgebied. Hoewel sommige belanghebbenden al maatregelen nemen om de waterverontreiniging onder controle te houden, is meer nadruk nodig om watervervuiling van verschillende punt- en niet-puntbronnen te monitoren en te beheersen. Hiervoor is het van cruciaal belang dat de monitoring frequentie van de waterkwaliteit wordt verhoogd en dat real-time waterkwaliteitssensoren op hotspotlocaties worden geïnstalleerd.
- Er is een goed meetnet voor de waterkwaliteit in Zuid-Afrika, maar er is een alarmerende trend van afname van de frequentie van het nemen van watermonsters (van wekelijks tot eens per maand of per kwartaal). Mozambique en Swaziland daarentegen moeten hun monitoringnetwerken verbeteren en kunnen profiteren van meer samenwerking met Zuid-Afrika.
- Routinematig verzamelde gegevens over de waterkwaliteit kunnen worden gebruikt voor het kalibreren van recursieve digitale filters, die op basis van dagelijkse afvoergegevens de componenten van baseflow kan schatten. Deze informatie kan nuttig zijn voor de operationele kwantificering van ecologische rivierafvoeren, voor operationeel waterbeheer, en voor het opzetten en verbeteren van hydrologische modellen in de regio.
- Hoewel tracerstudies erg duur kunnen zijn, met een goede aanpak kunnen ze goede inzichten bieden in de dynamiek van afvoerprocessen in stroomgebieden.
- Er is een sterke correlatie gevonden tussen antecedent neerslagindex en directe afvoer in het stroomgebied van de Kaap. Dit is waarschijnlijk ook het geval voor andere semi-aride subtropische stroomgebieden. Natte omstandigheden voorafgaand aan regenval vullen waterberging in het stroomgebied en leiden tot hogere bijdragen aan directe afvoer. Dit resulteert in een snellere reactie van het stroomgebied op regenvalgebeurtenissen.
- Landschapskartering, met behulp van benaderingen zoals HAND, is nuttig om meer informatie te halen uit beschikbare gegevens in het stroomgebied.

Verder zijn er verschillende recente bronnen van gegevens die enorm kunnen helpen bij het modelleren van grensoverschrijdende stroomgebieden, zoals de Soil grids 250m database van bodems, regenval gegevens ob basis van satellietbeelden (CHIRPS) alsmede schattingen van verdamping (ALEXI, CMERST, SSEBop, enz.). Het verfijnen van de ruimtelijke en temporele resolutie van deze producten zowel als biascorrectie moeten worden voortgezet als verbeterde inputs voor modellen.

- Het gebruik van de Soil grids 250m database leverde hydrologische modelleringsresultaten die net zo goed waren als het gebruik van lokale bodemgegevens - dit is dus een veelbelovende methode voor het modelleren van de Incomati rivier en andere stroomgebieden met gebrekkige data, aangezien dit een vrij beschikbare dataset is.

De belangrijkste aanbevelingen van deze studie zijn:

- Het versterken van meetnetten voor neerslag, afvoer, grondwater- en waterkwaliteit is noodzakelijk, met name in Mozambique en Swaziland. Het zou ook goed zijn als de landen databases delen en vergelijkbare protocollen volgen voor het verzamelen en rapporteren van gegevens.
- Het uitvoeren van een zorgvuldige en stroomgebiedswijde evaluatie van alle baten van het gebruik van water, en het opnieuw beoordelen welk water gebruik in het Incomati stroomgebied de hoogste prioriteit zou moeten hebben.
- Het valideren van de bevinding dat in semi-aride stroomgebieden de regionalisatie (transfer in ruimte) van recursieve digitale filterparameters mogelijk is met behulp van hoogfrequente waterkwaliteitsgegevens. Dit kan worden bereikt door de installatie van real-time EC-sensoren in geselecteerde substroomgebieden voor gedetailleerde kalibratie / validatie.
- Het uitvoeren van een uitgebreid onderzoek naar bias-correctie, verfijnen (downscaling) en kalibratie van regenval en actuele verdampingsgegevens gebaseerd op satellietbeelden, te gebruiken als input voor hydrologische modellering in semi-aride stroomgebieden. Dit is vooral relevant in grensoverschrijdende stroomgebieden zoals de Incomati, met onvergelijkbare datasets.

De belangrijkste wetenschappelijke innovaties van dit proefschrift zijn de toepassing van waterkwaliteitsgegevens om het begrip van afvoer-componenten in een semi-aride subtropisch stroomgebied te kwantificeren en te verbeteren. Verder is het testen van een methode om de recursieve digitale filters te kalibreren met behulp van direct beschikbare waterkwaliteitsgegevens een belangrijke stap in het verbeteren van de kwantificering van *baseflow*-componenten, nodig om ecologische rivierafvoeren te bepalen. Verder werden verschillende parameters geïdentificeerd voor hydrografische separatie in semi-aride gebieden, die bijdragen aan de wetenschappelijke kennis van dergelijke systemen. Parameters voor het STREAM-model toegepast op het Kaapstroomgebied dienen ook als maatstaf voor andere soortgelijke stroomgebieden, evenals het gevolgde proces om modelsimulaties te verbeteren door processtudies, landschapskartering en nieuwe gegevensbronnen te gebruiken.

Met het oog op de steeds maar toenemende druk op water is de maatschappelijke relevantie van dit proefschrift dat het een veelomvattende evaluatiemethode presenteert van de beschikbaarheid van watervoorraden en variabiliteit in de Incomati. Het recentelijk afgesloten *Progressive Realisation of the Inco-Maputo Agreement (PRIMA)* -project heeft verschillende IWRM-strategieën en -plannen voorgesteld om de uitdagingen van waterbeheer in de Incomati aan te pakken. De landen zijn van plan de wateropslagcapaciteit te vergroten door nieuwe dammen te bouwen en bestaande dammen te vergroten. Er zijn ook verschillende waterontwikkelingsprojecten gepland, waaronder de toename van de geïrrigeerde landbouw en commerciële bosbouw en de onttrekking van water voor stedelijk gebruik, inclusief voor de stad Maputo. De kennis van de temporele variabiliteit van watervoorraden en met name de bijdrage van grondwater aan de opbrengst van het systeem is echter niet goed begrepen. Dit probleem is nog urgenter wanneer rekening wordt gehouden met het plan om bijna-*real-time* ecologische rivierafvoeren te implementeren. Bovendien hadden eerdere studies een onvolledig inzicht in de afvoer-generatieprocessen in het stroomgebied. Dit onderzoek vult deze lacunes op door gebruik te maken van een combinatie van meerdere methoden om de hydrologische processen beter te begrijpen.

Verder helpt dit onderzoek om licht te werpen op *hotspots* waar de huidige wijzigingen in landgebruik de beschikbaarheid van waterbronnen beïnvloeden. Bovendien heeft het onderzoek verschillende studies in de regio beoordeeld en is het een goed startpunt voor degenen die onderzoek doen in de Incomati. Het testen van nieuwe methoden om optimaal gebruik te maken van beschikbare gegevens is van groot belang in deze regio, omdat waterbeheerders met beperkte datasets moeten werken; als zij de meest relevante informatie kunnen halen uit gegevens die routinematig verzamelde worden heeft dat grote toegevoegde waarde. In het bijzonder kan de kwantificering van afvoer-componenten, door middel van hydrografische separatie, nuttig zijn voor bepaling van ecologische rivierafvoeren en beheer van lage afvoeren gedurende de droge tijd.

Contents

1

GENERAL INTRODUCTION

1.1 Background

Water scarcity is a major issue in today's development agenda (Falkenmark, 1997; Savenije, 2000; Rijsberman, 2006; Davies and Simonovic, 2011). For sub-Saharan Africa in particular, several authors predict dramatic water shortages in the coming years (CAWMA, 2007; Davies and Simonovic, 2011) and this is likely to be exacerbated by the complications of managing river basins that cross international borders (Savenije and van der Zaag, 2000). The major issue in sharing an international water resources system is its utter scale and the opaqueness of system interactions over large distances (upstream and downstream). For example, it is difficult to attribute and quantify the consequences of upstream land use changes on downstream flood levels (Carmo Vaz and Lopes Pereira, 2000; Carmo Vaz and Van Der Zaag, 2003; Sengo *et al.*, 2005). This opaqueness may result in unexpected negative consequences of human interventions that are difficult to correct and may augment tensions between riparian countries sharing the basin's resources.

There are strong interdependencies in heavily committed basins with several diverse water use(r)s located in different riparian countries. Therefore, water allocation decisions have significant economic, social, environmental and political consequences. Often, decision-making involves difficult trade-offs. To assist, a number of decision-support tools have been developed (Jewitt and Görgens, 2000; DWAF, 2003b; Dlamini, 2007; DWAF, 2009a), most of which are based on optimising an economic objective function subject to constraints such as hydrological processes.

In the context described above, there is a clear requirement for integrated water resources management in order to balance food security, other economic needs and the needs of the environment (Molden, 1997; van der Zaag *et al.*, 2002; Rockström *et al.*, 2004). Such planning requires an understanding of the hydrological processes dominant in the catchment (Schulze, 2000; Uhlenbrook *et al.*, 2004; Lorentz *et al.*, 2008), and thus the factors that control the availability and vulnerability of (future) water resources (Uhlenbrook, 2003; Uhlenbrook, 2006; Uhlenbrook, 2009).

Hydrological processes at catchment scale are mainly dependant on climatic and physiographic controls such as rainfall, temperature, evaporation, soil characteristics, topography, geology and on land use changes. In southern Africa, the magnitude as well as spatial and temporal heterogeneity of water scarcity are often poorly understood (Butterworth *et al.*, 1999; Schulze, 2000; Mul, 2009; Love *et al.*, 2010b;

Warburton *et al.*, 2010). In the absence of rigorous experimental studies, runoff generation processes are also often poorly understood.

1.2 Problem definition

The Incomati is a stressed river basin in terms of water resources (van der Zaag *et al.*, 2002; ICMA, 2010). In South Africa, it is considered a closed basin[1] with water requirements higher than available water resources, particularly if the water requirements of Mozambique and the ecological reserve are considered (DWAF, 2009e; ICMA, 2010; Pollard and du Toit, 2011b; Riddell *et al.*, 2014b). The result of this is that the ecological reserve[2] is not met and the cross-border flows into Mozambique have on many occasions been less than what is specified in various international agreements (DWAF, 2009e; ICMA, 2010; Riddell *et al.*, 2014b).

However, there is much development pressure on the basin, and demands for water from different sectors are ever increasing. Despite the many decision support tools in place, there is still a need for tools that allow water managers to make decisions on water allocation in a transparent and equitable manner, considering trade-offs between water users and the best socio-economic value for water (van der Zaag *et al.*, 2002).

In order to have such tools in place, it is crucial to have a good understanding of the hydrology of the basin, and a good assessment of water availability and water uses. Many models (JIBS, 2001; DWAF, 2003a; Nkomo and van der Zaag, 2004; DWAF, 2009d) have been set up for the hydrology and water resources assessment, but some of them work in a stochastic way, based on historical data. This means that some hydrological processes are not fully understood, and hence are poorly represented in the models.

---

[1] A water resource system is "closed" when there is no usable water leaving the system other than that necessary to meet minimum instream and outflow requirements (Keller *et al.*, 1998). According to Falkenmark and Molden (2008), a river basin is termed closed when additional water commitments for domestic, industrial, agricultural, or environmental uses cannot be met during all or part/s of a year, while in an open basin more water can be allocated and diverted.

[2] Ecological reserve: a particular water quality and quantity to be set aside to protect the ecological functioning of aquatic ecosystems before water uses such as industry or agriculture can be authorised [National Water Act (No. 36 of 1998), South Africa]

The main knowledge gaps on the hydrology of the Incomati centre around understanding the dominating runoff generation processes including their space-time variability, groundwater flow, groundwater and surface water interactions, return flows and transmission losses in irrigation canals. The impact of land use changes on water availability constitutes another important knowledge gap, particularly concerning the impact on low flows. Integration between different models of hydrology and water resources and better input data are also a major concern of the Inkomati catchment management agency in South Africa (DWAF, 2009e; ICMA, 2010).

Therefore, the Incomati River basin, despite its small size, is an important transboundary river basin. Many users rely on its heavily utilized water resources. It also has a relatively dense network of climatic and hydrological stations (at least in the South African part). Therefore, it is a basin suited to address the research questions of this thesis.

This research aims at improving the hydrological understanding of the semi-arid river basin, by improving the understanding of runoff generation processes and using new data sources and methods.

1.3 Research objectives

The overall objective of the research is to improve the understanding of the hydrology of Incomati River basin in order to support optimal use of water resources in the face of socio-economic and environmental changes, constraints and uncertainties.

The specific objectives are:

1. To analyse the hydro-climatic variability and land use changes and trends in the Incomati basin, and establish the drivers of such trends and their implications for water management;
2. To describe spatio-temporal variability of water quality and environmental tracers in the Incomati, and infer hydrological process understanding from tracer patterns in the Incomati basin, as well as implications of observed patterns for water management;

3. To improve understanding of runoff generation processes and seasonality of stream flow using tracer methods and hydrochemistry at event scale and seasonal/annual scale; and
4. To test and apply an appropriate hydrological model for a selected sub-catchment, integrating historical, remotely sensed and field data, for the improvement of hydrological understanding of the basin.

1.4 Outline of the thesis

This thesis is structured in eight chapters that include the introduction (Chapter 1) and conclusions and recommendations (Chapter 8). The body chapters (Chapters 2 to 7) follow the specific objectives presented above.

Chapter 2 presents the main information of the study area, including the location, topography, climate, geology, soils, land use and land cover, water use, hydrology and past research work in hydrology and water resources of the Incomati basin.

Chapter 3 describes the drivers of spatial and temporal variability of streamflow in the Incomati basin (Saraiva Okello *et al.*, 2015). Statistical analysis of trends in streamflow and rainfall were conducted, and drivers for such trends were identified at the basin scale. Implications of the identified trends for water management are also discussed.

Chapter 4 presents the isotopic and hydrochemical river profile of the Incomati basin. Spatial and temporal variability of water quality in the basin is presented, and implications for hydrological processes understanding and water management are discussed.

Chapter 5 focuses on the use of tracers and digital filters to quantify runoff components in a selected sub catchment of the Incomati, the Kaap catchment (Saraiva Okello *et al.*, 2018b). The secondary water quality data is used to calibrate digital filters, and thus provide more comprehensive understanding of runoff components in the catchment.

Chapter 6 presents an experimental field analysis of water isotopes and natural tracers to determine the contribution of the different runoff components during a wet season in the Kaap catchment. Hydrograph separation using isotopes, and hydrochemistry is conducted for four main events captured during the wet season, and the dynamics of runoff generation processes are discussed (Camacho Suarez *et al.*, 2015).

Chapter 7 presents the set-up of the process-oriented hydrological model and the definition of the different runoff generation zones, based on landscape classification and information from process studies. The model is evaluated using runoff signatures, and gaps in process understanding and modelling are identified. The implications of the findings for water management are also (Saraiva Okello *et al.*, 2018a).

Chapter 8 synthesizes the main findings and proposes recommendations for better management of water resources and for future research.

The list of abbreviations together with a short biography of the author and list of publications can be found at the end of the book.

2

DESCRIPTION OF THE STUDY AREA

This chapter introduces the transboundary Incomati River basin. The location, physiographic and socio-economic characteristics of the basin are presented. A brief review of the hydrology, and past studies of water resources in the catchment are also presented. The RISKOMAN project, which was the umbrella under which the current PhD work took place, is also described.

2.1 Location and sub catchments

The Incomati[3] River basin is located in the south-eastern part of Africa and occupies portions of the Kingdom of Swaziland (Swaziland), the Republic of Mozambique (Mozambique) and the Republic of South Africa (South Africa) as shown in Figure 2.1. The total basin area is about 46 748 km^2 of which 2 561 km^2 (5.5%), 15 506 km^2 (33.2%) and 28 681 km^2 (61.3%) is in Swaziland, Mozambique and South Africa respectively (ICMA, 2010). The Incomati watercourse includes the Komati, Crocodile, Sabie, Massintoto, Uanetze and Mazimechopes Rivers and the estuary (TIA, 2002). The Komati, Crocodile and Sabie are the main sub-catchments, contributing about 94% of the natural discharge, with an area of 61% of the basin (Table 2.1).

Table 2-1. Sub-catchments of Incomati River basin, respective area and natural discharge[4]

Subcatchment	Catchment area	Natural discharge	
	$10^6 m^2$	$10^6 m^3 a^{-1}$	$mm\ a^{-1}$
Komati	11200	1420	127
Crocodile	10470	1226	117
Sabie / Sand	7050	750	106
Massintoto/ Nwaswitsontso	3430	22	6
Uanetze/ Nwanedzi	3930	14	4
Mazimchope	3970	21	5
Incomati*	6690	134	20
Total Incomati River Basin	**46748**	**3587**	**77**

Source: ICMA (2010); Sengo et al.(2005)
*Incomati sub-catchment is defined as the catchment area along the main stem inside Mozambique, after the confluence of Crocodile and Komati Rivers.

---

[3] The name "Incomati" is defined here as the basin that encompasses the river drainage region across the nations of Mozambique, South Africa and Swaziland. "Inkomati", spelt with a 'k', is the part of the basin within South Africa and Swaziland. In Mozambique "Incomati river" is used to refer to the main stem of the river inside that country.

[4] Natural discharge, also referred as virgin discharge (JIBS, 2001), is the term used to designate the discharge that would be generated under natural conditions of land cover and flow regime (without human intervention on land use, land cover and flow regulation, e.g. dams, afforestation, irrigation). The value is based on mean discharge.

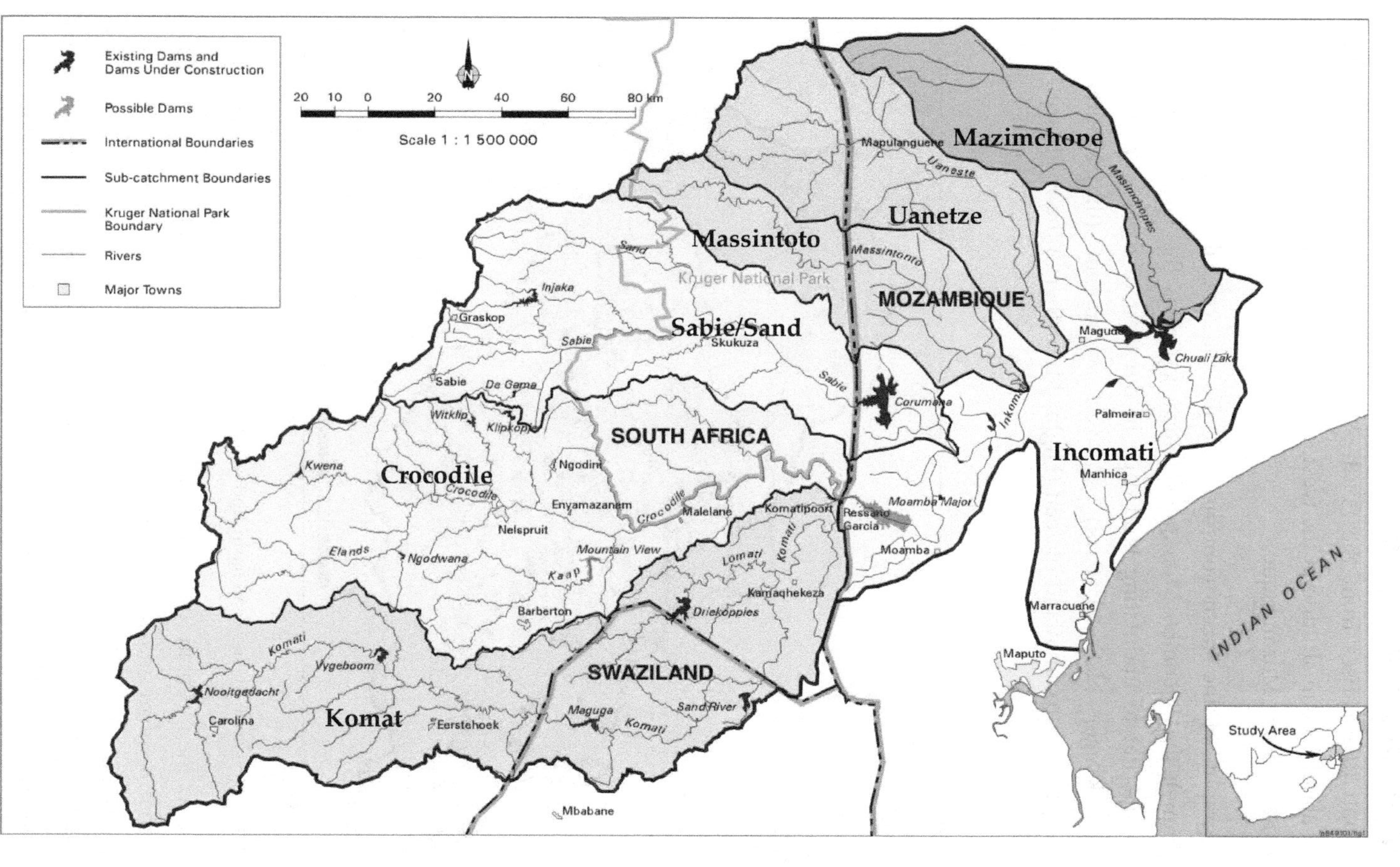

Figure 2-1. The Incomati River basin. Source: JIBS (2001).

2.2 Topography and climate

The Incomati River arises from the South African Highveld and the Transvaal plateau at about 2000 m altitude in the west of the basin and ends in the flat coastal plains near Maputo in Mozambique.

The topography of the basin comprises of the flat coastal plains in the east (Mozambique); the Lebombo Mountains which separates the Lowveld from the coastal plains; flat to undulating landscape in the west of the Lowveld (mostly within the Kruger National Park); and an escarpment (the Mpumalanga Drakensberg) rising to an inland plateau (Highveld) further to the west (see Figure 2.2).

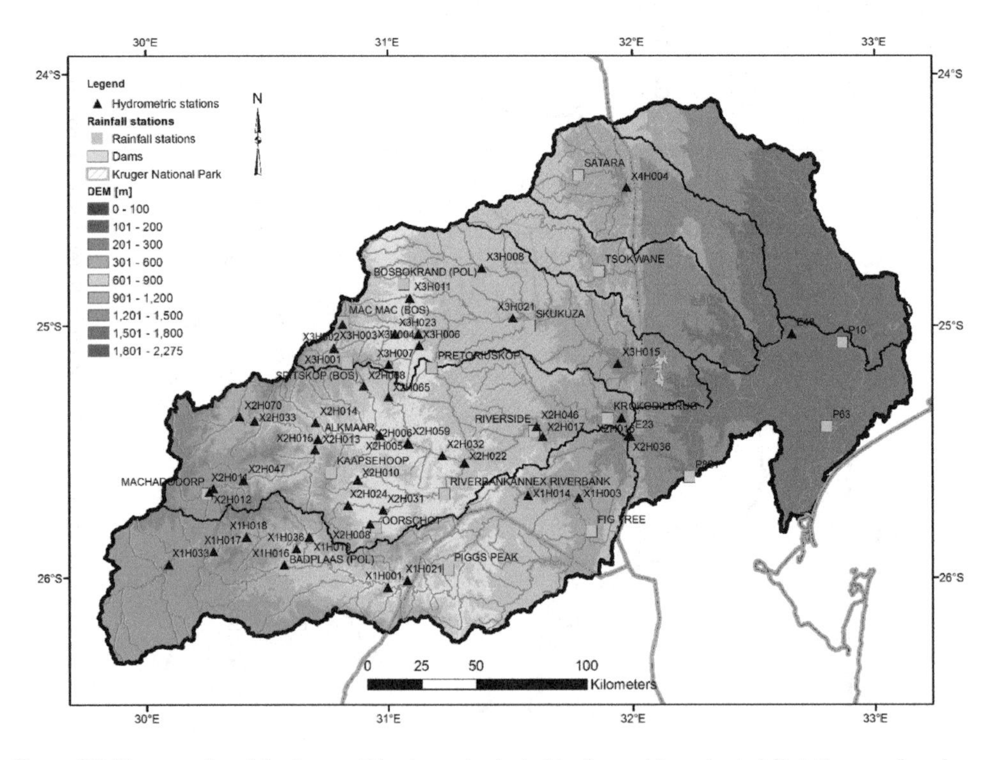

Figure 2-2. Topography of the Incomati basin, and selected hydrometric and rainfall stations analysed

The climate in the Incomati River basin follows the diverse geography of the basin, ranging from a warm and humid climate in the Mozambique coastal plain, to a cooler dry climate in the Highveld. The rainy season occurs between October and March, with tropical cyclones affecting mostly the lowlands. The mean annual precipitation is about 740 mm a^{-1}, whereas the mean annual potential evaporation is

1900 mm a^{-1}. However, while the precipitation increases from east to west, the evaporation decreases from east to west, resulting in an increasing deficit between rainfall and potential evaporation from west to east, and higher demands for irrigation towards the east (Carmo Vaz and Van Der Zaag, 2003; Sengo *et al.*, 2005). Figure 2.3 illustrates the mean monthly values for precipitation, temperature and potential evaporation for Nelspruit (on Crocodile catchment, west of the basin) and Satara (on Uanetze catchment, north of the basin; see Figure 2-2 for station locations).

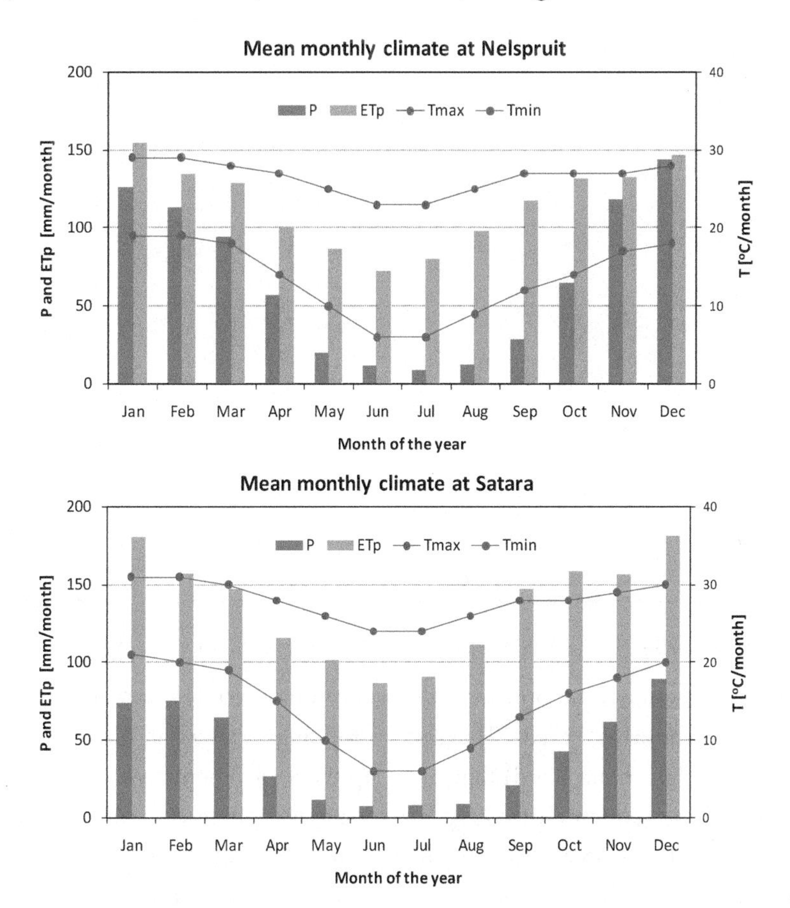

Figure 2-3. Mean monthly climate of the Incomati Basin, illustrated by precipitation (P), maximum and minimum temperature (T) and potential evapotranspiration (ETp) at the two climatic stations Nelspruit and Satara. Data source: Department of Water and Sanitation in South Africa (DWS).

2.3 Geology and soils

The geology of the basin is complex, characterized by sedimentary, volcanic, granitic, and dolomitic rocks, as well as quaternary and recent deposits (Figure 2.4).

There are occurrences of various minerals but only coal, asbestos, and gold are mined (Carmo Vaz and Van Der Zaag, 2003).

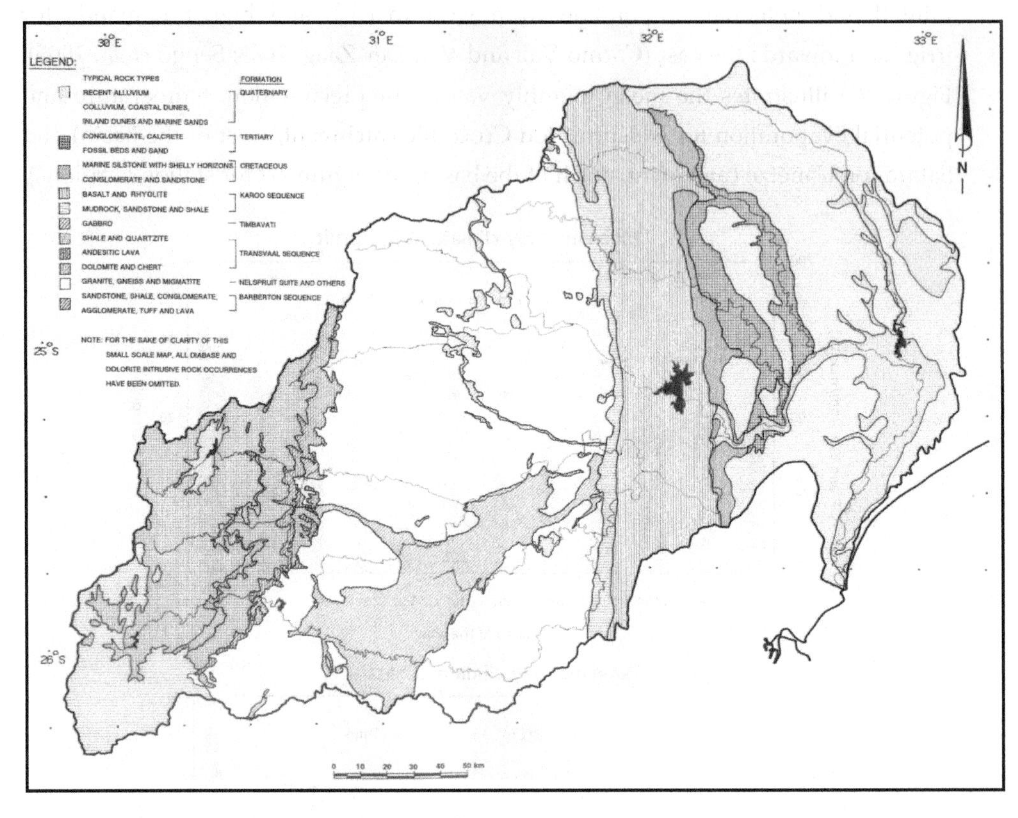

Figure 2-4. General geology of Incomati basin. Source: JIBS (2001)

The catchment is underlain on the western plateau by coarse sedimentary rocks and contains large endorheic[5] areas. These shifts to the quartzite and ancient greenstone belt (Barberton Mountain Land) that make up the escarpment areas and followed in the Lowveld by a sequence of younger extrusive igneous rocks such as granite and basalt. The coastal plain in Mozambique is underlain in the west adjacent to the Lebombo Mountains by basalt and thereafter to the east by fine sedimentary rocks and alluvium (JIBS, 2001).

The soils in the basin are highly variable. In the western part of the Komati River catchment there are occurrences of moderately deep clayey loam, with an

[5] Portion of a hydrological catchment that does not contribute towards river flow in its own catchment (local) or to river flow in downstream catchments (global).

undulating relief. Large parts of the Incomati (in South Africa) are covered by moderately deep sandy loam, with an undulating relief. Most of the central part of the Incomati consists of moderate to deep clayey loam with a steep relief. The eastern parts of the basin consist mostly of moderately deep clayey soils with an undulating relief (JIBS, 2001).

2.4 Land cover and land use

The basin is characterized by a wide variety of natural vegetation types. These vary between beaches and recent dunes, tropical bush and forest, and different types of savannah and grassveld (Carmo Vaz and Van Der Zaag, 2003).

The main vegetation classes found within the Incomati River basin are Lowveld, Bushveld in the lower mountains, Grasslands on the highveld, sourveld, thornveld, forest in patches against the escarpment and temperate freshwater wetlands in the upper Komati west of the town of Carolina (Acocks, 1988).

The dominant land uses in the catchment (Figure 2-5) are commercial forest plantations of exotic trees (pine, eucalyptus) in the escarpment region, dryland crops (maize) and grazing in the Highveld region and irrigated agriculture (sugarcane, vegetables and citrus) in the Lowveld (DWAF, 2009e; Riddell *et al.*, 2014b). In the Mozambican coastal plains, sugarcane and subsistence farming dominate. A substantial part of the basin has been declared a conservation area, which includes the recently established Greater Limpopo Transfrontier Park (the Kruger National Park in South Africa and the Limpopo National Park in Mozambique are part of it) (TPTC, 2010). It is also important to note that SANBI[6] identified the Incomati as a freshwater biodiversity hotspot in South Africa (ICMA, 2010).

[6] South African National Botanical Institute

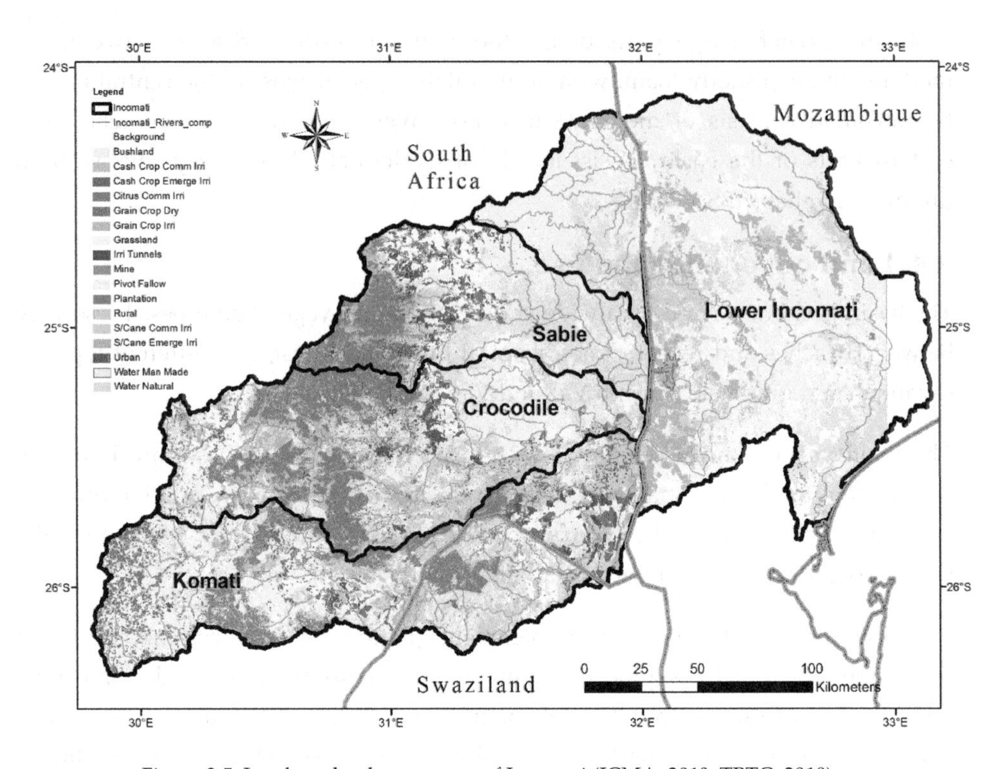

Figure 2-5. Land use land cover map of Incomati (ICMA, 2010; TPTC, 2010)

2.5 Water use, infrastructure and economy

The Incomati River basin is highly regulated. The level of water abstraction in the Incomati River is very high and the water demand is projected to increase in the future, as a result of further economic development and population growth (Nkomo and van der Zaag, 2004; LeMarie *et al.*, 2006; Pollard *et al.*, 2011). By the year 2002 total net consumptive water use was estimated at 2227 $x10^6m^3$ a^{-1} or 51% of the average amount of surface water generated in the basin (Table 2.2). The major water consumers (Table 2-2), accounting for 91% of all consumptive water uses, are the irrigation and forestry sectors, followed by inter-basin water transfers to the Umbeluzi Basin and the Olifants Catchment in the Limpopo Basin (Van der Zaag and Vaz, 2003; DWAF, 2009e; TPTC, 2010). From the late 1960s major dams (see Figure 2.1 for dam location) have been commissioned that have allowed increased water withdrawals at increasing levels of assurance (Table 2.3). All these developments have boosted the economies of the three riparian countries, but have also impacted on the environment (Sengo *et al.*, 2005). The area of irrigated agriculture and forestry has increased steadily, particularly in the Komati and Crocodile systems (Table 2-4).

Table 2-2. Summary of estimated natural streamflow, water demands in the Incomati Basin in 10^6 m^3 per year (TPTC, 2010)

	Natural MAR	First Priority Supplies*	Irrigation Supplies	Afforestration	Total Water Use
Komati	1,332	141.5	621	117	879.5
Crocodile	1,124	74.7	482	158	714.7
Sabie	668	30	98	90	218
Massintoto	41	0.3	0	0	0.3
Uanetse	33	0.3	0	0	0.3
Mazimechopes	20	0	0	0	0
Lower Incomati	258	1.5	412.8	0	414.3
Mozambique	325		412.8		
South Africa	2,663		961		
Swaziland	488		240		
Total	**3,476**	**248**	**1,614**	**365**	**2,227**

*First priority supplies include domestic and industrial uses

Table 2-3. Major dams (> $10 \times 10^6 m^3$) in the Incomati. Source: Carmo Vaz and Van Der Zaag (2003)

Tributary	Country	Major dam	Year commissioned	Storage capacity ($10^6 m^3$)
Komati	South Africa	Nooitgedacht	1962	81
Komati	South Africa	Vygeboom	1971	84
Komati	Swaziland	Maguga	2002	332
Komati	Swaziland	Sand River	1966	49
Lomati	South Africa	Driekoppies	1998	251
Crocodile	South Africa	Kwena	1984	155
Crocodile	South Africa	Witklip	1979	12
Crocodile	South Africa	Klipkopje	1979	12
Sabie	South Africa	Da Gama	1979	14
Sabie	South Africa	Injaka	2001	120
Sabie	Mozambique	Corumana	1988	879

Table 2-4. Summary Land use and water use change from 1950's to 2004 in Komati, Crocodile and Sabie sub-catchments. Source: adapted from (TPTC, 2010)

		1950's	1970's	1996	2004
Komati	Irrigation area (km^2)	17.6	144.1	385.1	512.4
	Afforested area (km^2)	247	377	661	801
	Domestic water use (10^6 m^{3}a^{-1})	0.5	7.7	15.5	19.7
	Industrial and mining water use (10^6 m^{3}a^{-1})	0	0	0.5	0.5
	Water Transfers out (10^6 m^{3}a^{-1}):				
	To Power stations in South Africa	3.4	103	98.1	104.7
	To irrigation in Swaziland outside Komati	0	111.8	122.2	121.8
Crocodile	Irrigation area (km^2)	92.8	365.8	427	510.7
	Afforested area (km^2)	375	1550	1811	1941
	Domestic water use (10^6 m^{3}a^{-1})	3	12.2	33.6	52.4
	Industrial and mining water use (10^6 m^{3}a^{-1})	0.1	7.5	19.8	22.3
Sabie	Irrigation area (km^2)	27.7	68.4	113.4	127.6
	Afforested area (km^2)	428	729	708	853
	Domestic water use (10^6 m^{3}a^{-1})	2.4	5.3	13	26.7
	Industrial and mining water use (10^6 m^{3}a^{-1})	0	0	0	0

2.6 Hydrology

2.6.1 Surface water

JIBS (2001) study estimated the total net natural runoff in the basin to be 3 587 x 10^6m^3 a^{-1}, of which 82% is generated in South Africa, 13% in Swaziland and 5% in Mozambique. About 80% of all runoff in a hydrological year (October–September) is produced during the months November–April. There are significant variations of discharge from year to year with floods and droughts occurring regularly, and the coefficient of variation of annual discharges is 50–65% (Carmo Vaz and Van Der Zaag, 2003; Sengo *et al.*, 2005). The average annual runoff at Ressano Garcia (border between South Africa and Mozambique) during the dry 1991-1995 period was only 12 percent of the long-term average measured over 1952–79. In the February 2000 floods, the Sabie River at Skukuza (catchment area 2 500 km^2) had a peak discharge of 3 500m^{3}s^{-1} (Smithers *et al.*, 2001). Table 2-2 shows the estimated natural runoff per sub-catchment.

An average of 150 t km^{-2}a^{-1} of soil is carried with the storm floods annually, occasionally increasing to 450 t km^{-2}a^{-1}, according to the JIBS (2001) estimate. Surface water quality is usually adequate for domestic and urban use after normal treatment. It is also suitable for irrigation. However, there is evidence of increasing quality degradation in some parts of the study basin (e.g. lower reaches of Komati, Crocodile and Sabie Rivers)(DWAF, 2009f).

Several models have been setup for hydrology and water resources simulation and assessment on Incomati, particularly in the South African sub-catchments (Komati, Crocodile and Sabie-Sand). Table 2-5 shows a list of reports of Hydrology and water resources and Table 2-6 lists some of the models implemented for different sub-catchments and areas of the Incomati.

2.6.2 Groundwater

Groundwater occurs in sufficient quantities for large-scale development only in the dolomites of the Transvaal Sequence, the Barberton Greenstone Belt, the alluvium of the Incomati River valley in the Mozambique coastal plain (with an estimated rate of recharge of about $150 \times 10^6 m^3 a^{-1}$), and in the Aeolian sands in the east of the Mozambique coastal plain (recharge is about $29 \times 10^6 m^3 a^{-1}$). The JIBS (2001) divided the Incomati River basin into hydrogeological regions based on the main aquifer types likely to occur within a lithological unit and the borehole yield potential.

Primary aquifers

With the exception of minor alluvial deposits along the river courses of the Lowveld Plains most of the primary aquifers occur within the Lower Incomati Coastal basin in Mozambique. Groundwater tends to be highly saline with the only exploitable water being hydraulically connected to the main river systems. The renewable resource of the alluvial systems is estimated at $60 \times 10^6 m^3 a^{-1}$ without risk of salinization [DNA, 1985 cited by (JIBS, 2001)]. The aeolian sands contain exploitable fresh water with an estimated recharge of $29 \times 10^6 m^3 a^{-1}$.

Secondary aquifers

Weathered and fractured aquifers in granites and basalts contain dispersed groundwater held mainly in fracture zones. Higher yields can be expected in areas with deeper weathering profiles. The Barberton Greenstone Belt has a high yield potential within the major fault systems of the belt and along the granite/greenstone contacts.

The rocks of the Transvaal Sequence and the lower Karoo Sequence are regarded as having low potential, although borehole data indicates that some high yields are obtainable in the rocks of the lower Karoo Sequence.

The groundwater in the fine grained Cretaceous sediments are highly saline and of low potential. The porous Tertiary calcrenites and limestones produce highly saline groundwater in high yielding boreholes with small pockets of useable fresh water. Water quality may improve with depth in some areas (JIBS, 2001).

Karsts may be extensively developed within the dolomites of the Transvaal Sequence with a recharge estimated at $22 \times 10^6 m^3 a^{-1}$. Exploitation is hampered by access and laborious exploration methods. Exploitation of this resource could affect the base flow in rivers transecting this aquifer system (JIBS, 2001).

Mussa *et al.* (2015) conducted a study of groundwater as potential emergency source to mitigate droughts in the Crocodile subcatchment. The study characterized drought severity in the Crocodile using meteorological and hydrological drought indicators. They used a groundwater model (MODFLOW) to simulate the recharge and impacts of various scenarios of worst drought conditions. The study estimated long term recharge in the Crocodile catchment of $77.9 \ mm \cdot a^{-1}$, which corresponds to 9% of mean annual rainfall. The study also concluded that groundwater can be used for drought mitigation without much adverse consequences to water levels in most of the catchment.

Bakhit (2014) conducted a similar investigation in the Lomati catchment (subcatchment of the Komati), and also found that there is potential to use groundwater particularly for emergency situations. However, discrepancies and uncertainties in the naturalized flow data are also reported, which highlights the need to conduct more comprehensive groundwater investigations in the catchment.

Table 2-5: Compiled reports of hydrology and water resources; adapted from Riddell and Jewitt (2010)

Report	Year	Publisher/Author	Basin(s)
Surface Water Resources of South Africa	1981	Water Research Commission	National, South Africa
Komati River Basin Development	1984	Henry Olivier & Ass., Chunnet, Fourie & Part.	Komati
Inkomati River Basin, Hydrology of Crocodile River	1985	Chunnet, Fourie & Part.	Crocodile
Komati River Basin Development. Driekoppies & Maguga Dams	1987	Henry Olivier & Ass., Chunnet, Fourie & Part.	Komati
Sabie River Catchment	1990	Department of Water Affairs	Sabie
Inkomati River Basin, Extended Hydrology of Crocodile River	1990	Chunnet, Fourie & Part.	Crocodile
Monografia Hidrografica do Rio Inkomati	1991	Consultec Lda	Incomati
Komati River Basin Development within Swaziland	1992	Sir Alexander Gibb & Part., Hunting Technical Services Ltd	Komati
Surface Water Resources of South Africa 1990	1994	Water Research Commission	National, South Africa
National Water Resources Strategy	1995	Dept. of Water Affairs and Forestry	National, South Africa
Maguga Dam Project	1997	Maguga Dam Joint Venture	Komati
Mozambique Country Situation Report - Water Resources	1998	Consultec Lda	National Mozambique
Joint Incomati Basin Study	2001	Consultec Lda, BKS Acres Ltd	Incomati
State-of-Rivers Report: Crocodile, Sabie – Sand and Olifants River Systems. A report of the River Health Programme.	2001	Water Research Commission	Crocodile, Sabie-Sand and Olifants
Komati and Mbuluzi Study	2003	Knight Piesold Ltd	Umbeluzi and Komati
Maputo, Umbeluzi and Inkomati River Basins: System Analysis Report for National Water Development Plans and Joint Water Resources Study.	2003	BKS Ltd, ARASUL	Inkomati, Maputo, Umbeluzi
Inkomati Management Area Internal Strategic Perspective study (ISP)	2004	Dept. of Water Affairs and Forestry	Inkomati
Undertake the validation and verification of registered water use in the Olifants and Inkomati Water Management Areas	2006	Dept. of Water Affairs and Forestry	Inkomati, Olifants
Quality Report for Komati Catchment Ecological Water Requirements Study	2006	Dept. of Water Affairs and Forestry	Komati
Water Resources of South Africa 2005	2008	Water Research Commission	National, South Africa
Inkomati Water Availability Assessment Study (IWAAS)	2009	Dept. of Water Affairs and Forestry	Inkomati
The Inkomati Catchment Management Strategy	2010	Inkomati Catchment Management Agency	Inkomati
PRIMA studies	2012	Tripartite Technical Committee (TPTC) between Moçambique, South Africa and Swaziland	Incomati

Table 2-6: Water resources and hydrology models used in the Incomati area; adapted from Riddell and Jewitt (2010)

Model	Author	Purpose	Implementer	Type	Catchment
Mike Floodwatch (Mike Basin, NAM)	DHI - (DHI South Africa)	Operations	ICMA	Integrated Hydrology	Crocodile
WReMP (Water Resources Modelling Platform)	SJ Mallory - IWR Water Resources	Operations	ICMA	Water Resources/Yield	Crocodile
ACRU (Agrohydrological Model)	RE Schulze et al - University of KZN	Hydrology Research	DWAF	Hydrology	Sand
			DHI - South Africa	Hydrology (-economics)	Crocodile
			University of KZN	Hydrology	Sabie
WAS (Water Accounting System)	N Bernade DWAF	Operations	KOBWA	Water Resources/Accounting	Komati
			DWAF/ICMA	Water Resources/Accounting	Sabie-Sand
ISM Model (Integrated information system)	Laudon & Laudon	Operations	KOBWA	Water Quality	Komati
WRYM (Water Resources Yield Model)	DWAF, Acres & BKS	Hydrology Assessments	DWAF	Water Resources/Yield	Inkomati
WR2005/SPATSIM (Integrated Water Resources of South Africa 2005)	Pitman, Hughes	Hydrology Hydrology Assessments	DWAF	Hydrology	South Africa
WRSM2000 (Water Resource Management System Model)	Pitman et al	Hydrology Assessments	DWAF	Hydrology	Inkomati
WRSM90 (Water Resource Management System Model)	Pitman & Kakebeeke	Hydrology Assessments	DWAF	Hydrology	Inkomati
ISIS (Hydrodynamic model)	CEH, UK	Operations	KOBWA	Hydrodynamic	Komati
WAFLEX	Savenije	Hydrology Research	UNESCO-IHE	Water resources	Komati
WEAP21 (Water Evaluation and Planning Model)	Stockholm Environment Institute (SEI)	Water resources management	UNESCO-IHE	Water resources and hydrology	Komati
SEBAL (Surface Energy Balance Algorithm for Land)	Bastiaanssen	Evapotranspiration and crop water use	WaterWatch	Energy balance, remote sensing	Komati

2.7 Past research on the hydrology and water resources of the Incomati River basin

There have been several assessments of the water availability in the Incomati. The first whole-basin assessment was delivered by the JIBS (2001) study as part of the international cooperation in the late 1990s in order to determine water availability for the whole of the Incomati basin. Afterwards there was a review of the water availability in the South African part of the Incomati, namely in the Inkomati Water Management Area (WMA) as part of the Internal Strategic Perspective study (ISP), by DWAF (2004). This was followed by a comprehensive reassessment of the water availability of the Inkomati WMA by DWAF (2009e) called the Inkomati Water Availability Assessment Study (IWAAS). The most recent study at basin scale was the Progressive Realisation of the Inco-Maputo Agreement (PRIMA, 2012). Despite the breadth of studies in the upper part of the basin (South Africa and Swaziland), much less studies have been carried out in the downstream part (Mozambique). These studies are briefly reviewed below.

2.7.1 Joint Incomati Basin Study (JIBS)

The JIBS study objective was to determine past, present and future use and availability of water in the Incomati River basin, the potential for development and the level of development as foreseen by each of the three basin states, and the most effective means of regulating the flow of rivers in the river basin, in order to improve assurance of water supply to existing land use and to adequately assure the water required for likely future land use scenarios. The study described the natural characteristics of the catchment, the water infrastructure, water resources (surface, groundwater and water quality), ecological water requirements, legal and institutional aspects, water resources development (dams) and system's yield analysis.

The JIBS study utilized the Pitman based WRSM90 hydrological model to simulate the hydrology of the whole Incomati basin. The Pitman model is a conceptual rainfall-runoff model, widely used in southern Africa. It is a monthly time-step model that has a relatively large number of parameters associated with components that represent the main hydrological processes (interception, surface runoff, soil moisture storage and runoff, groundwater recharge and discharge, evapotranspiration losses and routing) operating at sub-basin scales of 10–1 000s of km^2. It has been applied for water resources assessments studies WRSM90,

WR2000, WR2005 and WR2012 in South Africa, the latest with updated versions of the model. The model was setup for the entire river basin, including ungauged sub-catchments. The Mean Annual Run-off (MAR) was estimated for the entire basin with parameterized scenarios for net virgin runoff compared with a scenario anticipating maximum permitted forestry conditions (400 000 ha). The result for the period of simulation from 1921-1989 was a total net virgin MAR of 3 587 x $10^6 m^3$ compared to a MAR under forestry of 3 070 x $10^6 m^3$.

Given that 82%, 13% and 5% of the total MAR is produced in South Africa, Swaziland and Mozambique, respectively, the JIBS (2001) reveals the significant impact that upstream land-use in South Africa has on the runoff received downstream in Mozambique.

The JIBS study also identified ecologically sensitive areas, and provided an estimate of 5 $m^3 s^{-1}$ for estuarine freshwater requirement (EFR), based on the desktop Building Block Methodology. However, it advices that more in depth investigation and monitoring are required to better quantify EFR and account for its seasonal variation.

As a result of the system yield analysis and 12 scenario simulations, the study cautions that future water requirements of 2694 x $10^6 m^3 a^{-1}$ (1107 x $10^6 m^3 a^{-1}$ was the present water requirement in 1991) can only be supplied with present infrastructure at very low assurances of supply, which are considered unacceptable. And, even with all planned new infrastructure in place (building of Moamba Major dam, Mountain View dam and raising of Corrumana dam), only small improvements in assurance of supply are achieved. Therefore, it recommends that the countries need to carefully prioritize their development plans, considering that water is a major constraint to development. Furthermore, it urges the start and intensification of baseline studies to better characterize instream flow requirements and estuarine freshwater requirements.

2.7.2 Inkomati Management Area Internal Strategic Perspective study (ISP)

The ISP utilized water availability assessment methodologies that are outlined in the National Water Resource Strategy of South Africa. The study determined available gross yield for the catchments in the Inkomati WMA minus uses for environmental reserve and stream flow reduction activity by alien invasive plants (DWAF, 2004; Riddell and Jewitt, 2010).

For the Komati sub-catchment the key issue identified was that there is insufficient water available to accommodate any additional allocations whilst at the same time there is a huge demand for additional water-use licenses. Crocodile and Sabie sub-catchments were dominated by irrigation and forestry which are by far the largest users of water in the catchment. However, Crocodile faces the challenge of underdeveloped water management infrastructure (only has one major dam - Kwena Dam) to cope with the demand.

The study estimated that groundwater mean annual recharge varies from 100 to 150 mm a^{-1} in the elevated higher rainfall areas along the western boundary and in the south of the basin to 10 to 20 mm a^{-1} in its low rainfall, much lower standing easternmost portion. These annual groundwater recharge values are equivalent to rates of about 120 000 m^3 km^{-2} and about 10000 m^3 km^{-2} to 20 000 m^3 km^{-2}, respectively. The study also estimated that registered groundwater use constituted only 2.8% of estimated annual recharge, and in general it would seem that groundwater is underutilized (DWAF, 2004).

2.7.3 Inkomati Water Availability Assessment Study (IWAAS)

The IWAAS study (DWAF, 2009e), which is the most updated hydrological study of the Inkomati (South Africa), sought to determine absolutely and comprehensively the water situation in the Inkomati WMA. The assessment was achieved through more precisely estimating the hydrology of the catchments within the Inkomati WMA using the updated Pitman hydrology model WRSM2000 at the quinary[7] level. And, then using the water resources yield model (WRYM) to estimate the actual water use of the catchment based on the naturalized hydrology of the catchment and the storage characteristics of dams in the system.

One of the main findings of the IWAAS study was significant reductions in the runoff of the catchments (DWAF, 2009e). The reduction in Mean Annual Run-off

---

[7] South Africa is divided into 22 primary drainage regions , Incomati being the region X. These are further subdivided into secondary (eg. X1-Komati, X2-Crocodile, X3 - Sabie), tertiary and quaternary sub-catchment areas. The quaternary sub-catchment is defined as the basic unit of water resources management. The Incomati region has about 104 quaternary catchments, with average area of 650 km^2. The quinary level is a further subdivision of quaternaries.

(MAR) on previous estimates was up by 5% in the Komati, between 7-14% and 10-13%, in the Crocodile catchment and the Sabie catchment respectively. Despite the reductions in MAR, it was suggested that the hydrology had not deviated greatly from previous studies.

The IWAAS study identified that the largest water user in the Inkomati WMA was indeed the irrigation sector and this was estimated using the irrigation component of the Water Quality Model. Lawful use was estimated in the basis of satellite imagery and the model crop water requirements.

Forestry has appeared to increase in the Inkomati WMA in recent years although very few licenses have been awarded. IWAAS study suggested that users are still receiving water at acceptable assurance levels, and this was attributed in large part to the operations of the new large dams in the catchment.

2.7.4 Interim Inco-Maputo Agreement (IIMA) and Progressive Realisation of the Inco-Maputo Agreement (PRIMA)

The Governments of the Republics of Mozambique, South Africa and the Kingdom of Swaziland have been working together on the management of their shared water resources and on carrying out joint studies of common interest through the Tripartite Permanent Technical Committee (TPTC), which was established in 1983. In 2002 the three governments signed an Interim IncoMaputo Agreement (IIMA) (TIA, 2002) and it was agreed that a Comprehensive Agreement would follow. This would allow the countries to more effectively utilise, develop and protect the shared waters of the Incomati and Maputo River Basins. The TPTC has, through a dedicated Task Group, developed an Implementation Activity and Action Plan (IAAP) which resulted in the identification of twelve projects:

- IAAP 1 – Shared Water Course Institutions,
- IAAP 2 – Review of National Water Policies and Legislation.
- IAAP 3 - Integrated Water Resource Management (IWRM).
- IAAP 4 - Augmentation of Water Supply to Maputo City and Metropolitan area.
- IAAP 5 - Disaster Management in the Incomati and Maputo Watercourses.
- IAAP 6 - Transboundary Water Quality Impacts.
- IAAP 7 - Exchange of and Access to Information.
- IAAP 8 - Capacity and Confidence Building.
- IAAP 9 - Stakeholder Participation and Communication.

- IAAP 10 - System Operating Rules.
- IAAP 11 - Process for the Comprehensive Agreement.
- IAAP 12 – Management of the IAAPs

These projects were implemented under the Progressive Realisation of the IncoMaputo Agreement (PRIMA), financed by the Government of Netherlands, from January 2007 to December 2011 (Aurecon, 2010; ICMA, 2010; Riddell and Jewitt, 2010). However, due to financial constraints only some of the IAAP were actually implemented (1, 2, 3, 4, 7, 9, 10 and 12).

IAAP 3 presents the most comprehensive review of the status quo of the Incomati basin, as well as the current and future water resources management situation and scenarios. An extensive consultation process was followed to identify common vision and scenarios of integrated water management in the basin, with the 3 countries representative stakeholders. A final scenario was chosen, and implementation plans developed accordingly. The study reinstated the fact that the Incomati basin is very stressed, with allocations above system yield in the Crocodile and Komati catchments. Furthermore, it highlighted that the environmental flow requirement were only based on desktop studies for locations in the Mozambican part of the basin and in the estuary – therefore further investigations are recommended.

2.7.5 Other water resources studies

Nkomo and van der Zaag (2004) developed a relatively simple, spreadsheet-based water resources model WAFLEX (Savenije, 1995) to analyse water availability and use under current and future scenarios, considering the Tripartite Interim Agreement (TIA, 2002), between the riparian countries. This study shows that the future water demands will result in appreciable shortages for irrigation and domestic use. The agreed maximum development levels will soon outstrip the ability of the catchment's supply. So in the Incomati, negotiations about water sharing could focus on the relative values of green and blue water resources (van der Zaag et al., 2002).

Sengo *et al.* (2005) studied the impact of alteration of the flow regime into the estuary, as a result of upstream developments in the basin, using statistical methods and the WAFLEX model. The results showed a significant decrease in the frequency of small flood events into the Incomati estuary for the period of analysis (1957 to 2001): the median (2 year) flood decreased from 625 x $10^6 m^3$ $month^{-1}$ under natural conditions to 440 x 10^6 m^3 $month^{-1}$ in the current 2002 situation (Sengo *et al.*, 2005). They concluded that the reduction of natural fresh water fluxes (small floods) has a negative impact on the state of the environment and hence on the goods and services the estuary yields.

Dlamini (2007) describes Decision Support Systems (DSS) developed by Komati Water Basin Authority (KOBWA) to manage the Komati system. These include DSS for: (i) long-term water allocation (yield model) between the countries, (ii) short-term water allocation (rationing model), and (iii) day-to-day water release (hydraulics model). An extensive water-monitoring program has been put in place to improve the effectiveness of these DSSs. Dlamini (2007) reports that there has been a wide acceptance of the DSS by the users in the basin since they enable water users and water managers to make transparent water management decisions.

Hellegers et al. (2009) proposed a method that combines remote sensing and socio-economic analyses to assess spatial variation in crop water productivity (CWP) and economic water productivity (EWP) in the Inkomati basin. The study compared actual water consumption, crop water productivity and economic water productivity of bananas, commercial and emergent sugarcane in the Lower Komati, showing that the reallocation of water from crops with lower CWP (e.g. sugarcane) to higher CWP (banana) is not the most effective economically. The method allows assessment of the size of potential gains and losses of spatial water reallocations. Hellegers et al. (2010) further expands on the method, and provides an example of how remote sensing estimates of ETact, ETpot, biomass production and rainfall can be used for water resources monitoring, allocation planning and determining the potential for water transfers through mechanisms such as water trading. The remote sensing data shows where water is being consumed, how this relates to assigned water rights, how land use affects water availability, and where water is being most and least productively used.

De Lange *et al.* (2010) developed an approach for overcoming spatial incompatibilities between socio-economic and biophysical data and applied to the irrigation agriculture sector in the Inkomati, in South Africa. This method allows integration of socio-economic and biophysical data to support water allocations within river basins, based on a meta-modelling approach using GIS and an application of a water-use simulation model.

Hughes and Mallory (2008) developed an approach and software to incorporate environmental flows into water resource management operation, in the South African water resources legislation context. The method developed relies upon simulating natural flow conditions based on inputs of near real-time observations of rainfall and a set of operating rules, with a focus on managing the variability of continuous low flows. They found that the main limitation to the successful implementation of the low flow approach is the lack of legislated control over run-of-river water abstractions.

Pollard and du Toit (2011b) and Riddell *et al.* (2014b) conducted research about the sustainability of the easterly-flowing rivers of the Lowveld of South Africa (Luvuvhu, Letaba, Olifants, Sabie-Sand, Crocodile and Komati Rivers). They assessed the state of compliance with the Ecological Reserve (ER), as well the challenges associated with an assessment of compliance. The ER in South Africa is defined as a function of the natural flow. However, because the natural flow in a system is not continuously available, it is generating issues with real-time implementation, Real-time hydrological models with accurate daily rainfall are required to estimate the natural flow, but such rainfall data are missing in many catchments (Pollard and du Toit, 2011b). The other problem identified is that water users, particularly irrigators, would like to know in advance their actual water allocation for the next growing season. Current water resources models can provide estimates of available water in the short-term but are unable to indicate how much water will be required for the ER due to uncertainties with the future flow (Pollard and du Toit, 2011b).

Riddell *et al.* (2014b) developed a semi-quantitative method to assess the compliance of the Crocodile River with the reserve in an historical context, based on flow-duration curve (FDC) analysis. They used both daily and monthly average flow data to quantify the extent and magnitude of non-compliant flows against environmental water requirements (EWRs) for three periods (1960-1983, 1983-2000, and 2000-2010). They found increasing degree of non-compliance with the reserve for each of these periods (14%, 35%, and 39% of the time), respectively, where effects were most prominent in the low-flow season.

Several studies were conducted to analyse the implementation of IWRM in the Incomati Basin (Slinger *et al.*, 2010; Gallego-Ayala and Juízo, 2012; Gallego-Ayala and Juízo, 2014). Slinger *et al.* (2010) analysed information use for decision making by the TPTC, and provide a useful review of how decision making in water management has evolved in the Incomati Basin. They identified several problems the countries face regarding transboundary water management, such as cultural and language differences, diversity in perception, inadequacy of stakeholder involvement, variability in political commitment, lack of capacity, lack of operational experience, the weak mandate of the international decision-making body, and the tense South African–Mozambican relationship. They recommended some changes for the successful implementation of IIMA, and highlighted the importance of the socio-political interface in the decision making process. Gallego-Ayala and Juízo (2012) tested a framework to evaluate river basin organization's performance in implementing IWRM, using the Incomati as one of the case studies. Gallego-Ayala and Juízo (2014) provides a tool to facilitate stakeholder's participation and involvement in planning and decision making, by using Analytical Hierarchy Process. They interviewed several stakeholders with competing interests in the Lower Incomati, to identify their preferences in terms of objectives and management plans for the catchment. The environmental sustainability plan was chosen as the preferred one by stakeholders, which again highlights the relevance of better understanding and quantifying environmental flows in the basin.

2.7.6 RISKOMAN and WATPLAN

Risk Based Operational Water Management for the Incomati Basin (RISKOMAN) is a jointly funded research project by UPaRF (The Netherlands, administered by IHE Delft) and the Water Research Commission (South Africa, administered by Centre for Water Resources Research of the University of KwaZulu-Natal (CWRR-UKZN)). The project aims at the development of a tool that would be channelled through the Inkomati Catchment Management Agency of South Africa (ICMA), Komati Water Basin Authority (KOBWA) in Swaziland and South Regional Water Administration (ARA-Sul) in Mozambique. The tool employs data on improved hydrological understanding of the Incomati through the use of new and innovative technologies used in the hydrological sciences to assist in the water allocation problem for this water stressed basin.

RISKOMAN-WRC is focused on improving the hydrological processes understanding in the basin particularly by incorporating remote sensing technologies to augment present and future modelling requirements. Furthermore, through this approach and social-learning interaction with decision-makers in the catchment it aims to develop a simple to use but conceptually and scientifically robust decision support tool and methodologies for Integrated Water Resources Management. Riddell *et al.* (2014a) summarizes key findings and results of the project. Jackson (2014) describes the operational tool developed and tested in the ICMA.

RISKOMAN- IHE is focused on water management issues such as water allocation and accounting, particularly with respect to water trade-offs between sectors, in an uncertain climatic regime, based on a sound hydrology understanding. Figure 2-6 shows the main activities of the RISKOMAN project. This PhD contributes mainly into the activity 1. The project yielded several outputs and publications, which can be found in https://riskoman.un-ihe.org.

WATPLAN project is a joint endeavour between WaterWatch in the Netherlands and CWRR-UKZN and is using new remote sensing (e.g. SEBAL) technologies to understanding basin scale water fluxes, most notably that of evapotranspiration, and the Incomati is the key study site for this project. The final report of the study is available at https://cordis.europa.eu/result/rcn/144038_en.html and key findings are documented in van Eekelen *et al.* (2015).

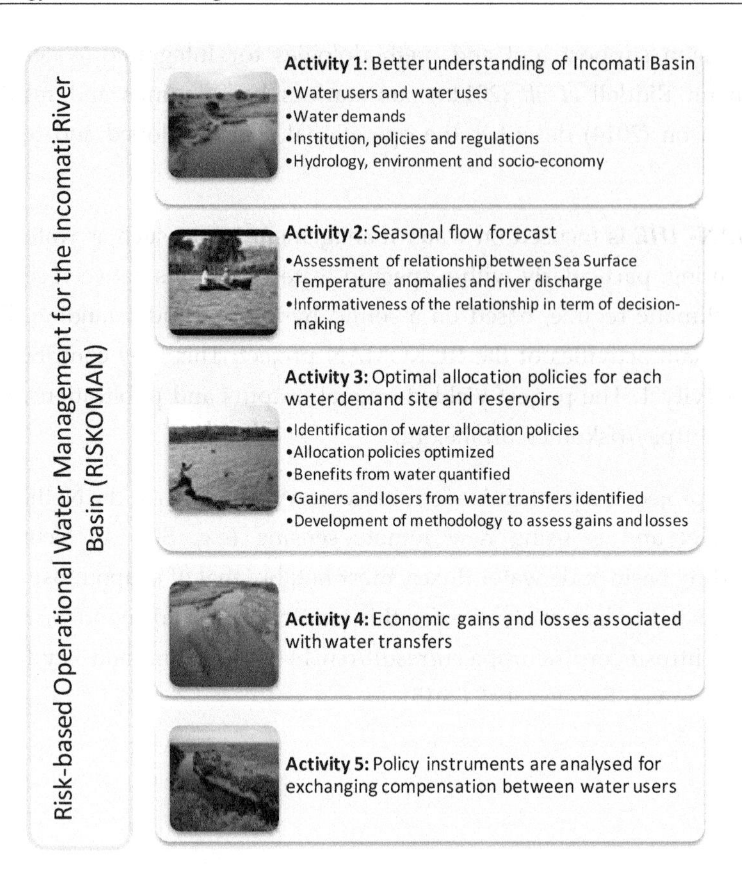

Figure 2-6. RISKOMAN project activities

2.8 Conclusion

Given the above review, and the wealth of previous research and gaps identified, the Incomati basin is an ideal setting to investigate the usefulness of alternative data sets, namely water quality and remote sensing, and new methods to improve hydrological process understanding, for improved water resource management. The following is the first of the substantive chapters of this study, and it focuses on understanding the hydrological variability of the Incomati basin, and the drivers of this variability.

3

DRIVERS OF SPATIAL AND TEMPORAL VARIABILITY OF STREAMFLOW IN THE INCOMATI RIVER BASIN

This chapter investigates the spatial and temporal variability of streamflow and rainfall in the Incomati basin. Long term rainfall records were analysed using descriptive statistics, annual anomalies, and trends. Change points were identified using the Pettitt test, and monthly and annual trends were identified using the Spearman test. Significance of change points and trends were assessed with F and t tests. Furthermore, the Indicators of Hydrological Alteration approach was used to further describe the streamflow regime in selected streamflow gauges in the catchment, and identify trends in key hydrological indicators. The trends were mapped across the basin, and were compared with rainfall trends, as well as historical and current land use. The links between observed changes and potential drivers were discussed, and well as the implications of this for water management in the basin.

This chapter is based on: Saraiva Okello, A. M. L., Masih, I., Uhlenbrook, S., Jewitt, G. P. W., van der Zaag, P., and Riddell, E., 2015: Drivers of spatial and temporal variability of streamflow in the Incomati River basin, *Hydrol. Earth Syst. Sci.*, 19, 657-673, 10.5194/hess-19-657-2015.

3.1 Introduction

Global changes, such as climate change, population growth, urbanization, industrial development and the expansion of agriculture, put huge pressure on natural resources, particularly water (Jewitt, 2006a; Milly *et al.*, 2008; Vörösmarty *et al.*, 2010; Miao *et al.*, 2012; Montanari *et al.*, 2013). In order to manage water in a sustainable manner, it is important to have a sound understanding of the processes that control its existence, the variability in time and space and our ability to quantify that variability (Jewitt *et al.*, 2004; Hu *et al.*, 2011; Montanari *et al.*, 2013; Hughes *et al.*, 2014).

Water is critically important to the economies and social well-being of the predominantly rural populations of southern Africa, where environmental sustainability issues are increasingly coming into conflict with human development objectives and where data are also scarce. The local economies and livelihoods of many southern African communities are strongly dependent on agriculture and fisheries, and water availability remains one of the main constraints to development in Africa (Jewitt, 2006a; Pollard and du Toit, 2009). Hydro-power is also locally important, while a substantial amount of foreign income is derived from wildlife tourism in some countries of the region (Hughes *et al.*, 2014).

Climate change intensifies the global hydrological cycle, leading to more frequent and variable extremes. For southern Africa, recent studies forecast an increase in the occurrence of drought due to decreased rainfall events (Shongwe *et al.*, 2009; Rouault *et al.*, 2010; Lennard *et al.*, 2013). Furthermore, it is expected that temperatures will rise, and thus the hydrological processes driven by it will intensify (Kruger and Shongwe, 2004; Schulze, 2011). Compounding the effect of climate change are the increased pressures on land and water use, owing to increased population and the consequent requirements for food, fuel and fibre (Rockström *et al.*, 2009; Warburton *et al.*, 2010; Warburton *et al.*, 2012). Areas of irrigated agriculture and forestry have been expanding steadily over the past decades. Urbanization also brings with it an increase in impervious areas and the increased abstraction of water for domestic, municipal and industrial purposes (Schulze, 2011).

In southern Africa, these pressures have led to changes in natural streamflow patterns. However, not many studies are available concerning the magnitude of such changes and what the main drivers are (Hughes *et al.*, 2014). Projections of the impact of climate change on the water resources of South Africa were investigated by Schulze (2012) and streamflow trends of some southern Africa rivers have been analysed (Fanta *et al.*, 2001; Love *et al.*, 2010b), but no such studies are available for the Incomati Basin.

The Incomati is a semi-arid trans-boundary river basin in southern Africa, which is water-stressed because of highly competing demands from, amongst others, irrigated agriculture, forestry, energy, environmental flow and basic human needs (DWAF, 2009e; TPTC, 2010). The impact of these demands, relative to the natural flow regime, is significant. Hence, there is an opportunity to improve water management, if a better scientific understanding of water resources availability and variability can be provided (Jewitt, 2006a).

The goal of this paper is to determine whether or not there have been significant changes in rainfall and streamflow during the time of record, and what the potential reasons and implications of such changes are. The main research questions are:

- Does the analysis of precipitation and streamflow records reveal any persistent trends?
- What are the drivers of these trends?
- What are the implications of these trends for water management?

The variability and changes of rainfall and streamflow records were analysed and the possible drivers of changes were identified from the literature. The spatial variation of trends on streamflow and their possible linkages with the main drivers are analysed. Based on the findings, approaches and alternatives for improved water resources management and planning are proposed.

3.2 Methodology

3.2.1 Study area

The Incomati River basin's location, physiographic and socio-economic characteristics are presented on Chapter 2, sections 2.1, 2.2, 2.3, 2.4, 2.5 and 2.6.

3.2.2 Data and Analysis

3.2.2.1 Rainfall

Annual, monthly and daily rainfall data for Southern Africa for the period of 1905 to 2000 was extracted from the Lynch (2003) database. The database consists of daily precipitation records for over 12000 stations in Southern Africa, and data quality was checked and some data were patched. The main custodians of rainfall data are SAWS (South Africa Weather Service), SASRI (South Africa Sugarcane Research Institute) and ISCW (Institute for Soil Climate and Water). About 20 stations out of 374 available for Incomati were selected for detailed analysis (Figure 2-2). The selection criteria were the quality of data, evaluated by the percentage of reliable data in the database, and the representative spatial coverage of the basin. Eight of the 20 stations' time series were extended up to 2012, using new data collected from the SAWS.

The spatial and temporal heterogeneity in rainfall across the study area was characterised using statistical analysis and annual anomalies. The time series of annual and monthly rainfall from each station was subjected to the Spearman Test, in order to identify trends for the period of 1950-2000 and 1950-2011. Two intersecting periods were chosen, to evaluate the consistency of the trends. Due to natural climatic variability, there are sequences of wetter and drier periods, so some trends appearing in a specific period might be absent when a longer or shorter period is considered. The Pettitt Test (Pettitt, 1979) is used to detect abrupt changes in the time series. Potential change points divide the time series in two sub-series. Then the significance of change of mean and variance of the two sub-series is evaluated by F and T-tests. Potential change points were evaluated with a 0.8 probability threshold and significance of change was assessed with F and T-test at 95% confidence level. (Zhang *et al.*, 2008; Love *et al.*, 2010b). The annual and monthly time series were also analysed for the presence of serial correlation. Tests were carried out using SPELL-stat v.1.5.1.0B (Guzman and Chu, 2004).

3.2.2.2 Streamflow

Streamflow data for 104 gauging stations in South Africa (obtained from the Department of Water and Sanitation DWS) and two gauging stations in Mozambique (obtained from ARA-Sul) were used in this study. Long time series of flow data were not available in Swaziland. Based on the quality of data, time series length, influence of infrastructure (dams, canals) and spatial distribution, 33 stations were selected for detailed analysis (Table 3-1 and Figure 2-2). As this catchment is highly modified, very few stations could be considered not impacted by human interventions. Data from pristine catchments can reveal the dynamics of natural variability of streamflow, and isolate the impacts of climate change on streamflow. An analysis of the indicators of hydrologic alteration was conducted, to identify patterns and trends of the streamflow record (a single period analysis for the entire time series and for the period of 1970-2011), as well as to assess the impact of infrastructure on the streamflow (two-period analysis, before and after the major infrastructure development).

3.2.2.3 Indicators of Hydrologic Alteration

The US Nature Conservancy developed a statistical software program known as the "Indicators of Hydrologic Alteration" (IHA) for assessing the degree to which human activities have changed flow regimes. The IHA method (Richter *et al.*, 1996; Richter *et al.*, 2003; Richter and Thomas, 2007) is based upon the concept that hydrologic regimes can be characterized by five ecologically-relevant attributes, listed in Table 3-2: (1) magnitude of monthly flow conditions; (2) magnitude and duration of extreme flow events (e.g. high and low flows); (3) the timing of extreme flow events; (4) frequency and duration of high and low flow pulses; and (5) the rate and frequency of changes in flows. It consists of 67 parameters, which are subdivided into two groups; 33 IHA parameters and 34 Environmental Flow Component parameters. These hydrologic parameters were developed based on their ecological relevance and their ability to reflect human-induced changes in flow regimes across a broad range of influences including dam operations, water diversions, ground-water pumping, and landscape modification (Mathews and Richter, 2007). 33 selected gauging stations from the Incomati Basin were analysed with this method using daily flow data. Many studies successfully applied the methodology of "Indicators of Hydrologic Alteration", in order to assess impacts on streamflow caused by anthropogenic drivers (Maingi and Marsh, 2002; Taylor *et al.*, 2003; Mathews and Richter, 2007; De Winnaar and Jewitt, 2010; Masih *et al.*, 2011). In the case of the present study, the indicators of magnitude of monthly flow, magnitude and duration of extreme flow, as well as timing were analysed for the period 1970-2011, to assess whether consistent trends of increase or decrease of the hydrological indicators were present.

The IHA software was used to identify trends in streamflow time series, based on the regression of least squares. This trend is evaluated with the P value, and only trends with $P \leq 0.05$ were considered significant. The value of the slope of the trend line indicating increasing or decreasing trend. This information was compiled and mapped for the various hydrological indicators using ArcGIS 9.3.

3.2.2.4 Land use analysis

Land use was analysed, based on secondary data, as remote sensing maps are only available after the current forestry plantations were already established. Additionally, a map of current land use (2011) (Jarmain *et al.*, 2013) and land use of 2000 (Van den Berg *et al.*, 2008) were compared with the maps of trends of indicators of hydrologic alteration. Where the occurrence of trends in flow regime was consistent with the changes in land use, the temporal evolution of the land use changes were further investigated.

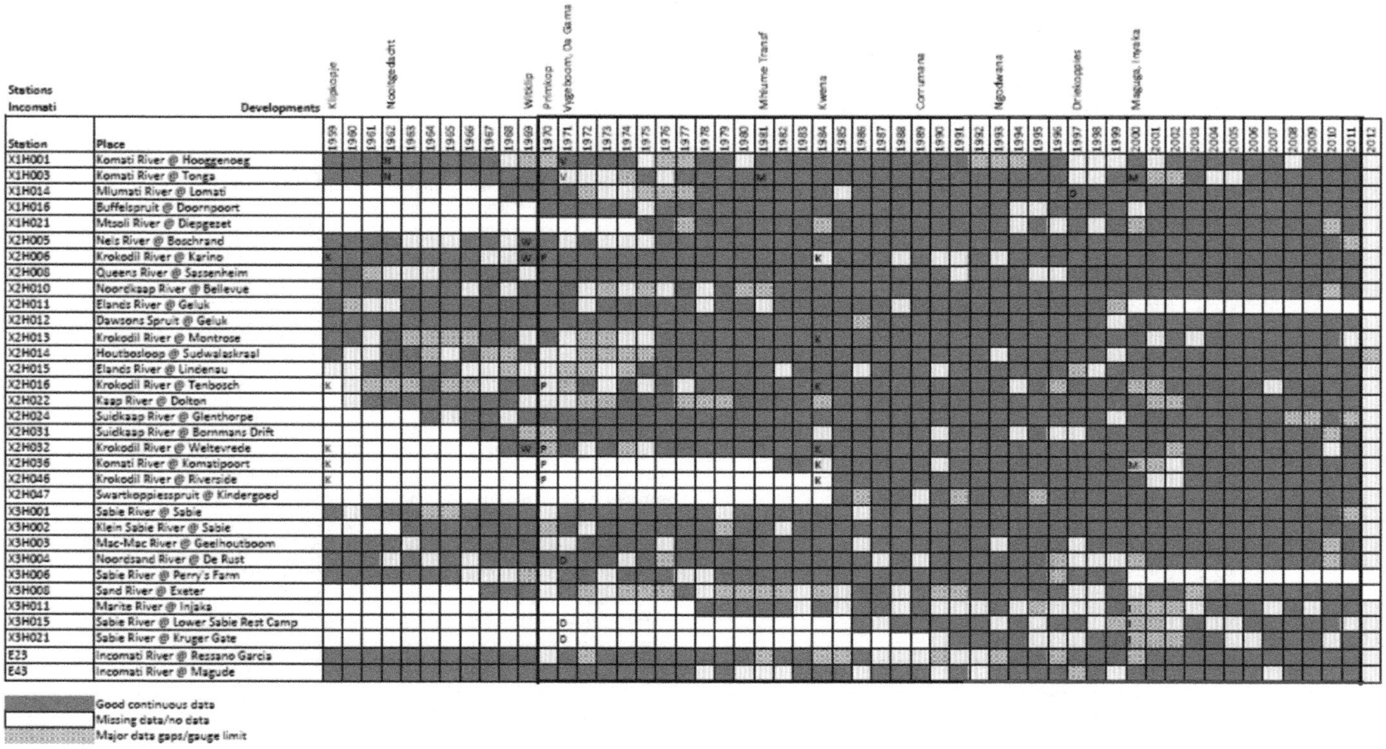

Stations
Incomati

Developments: Klipkopje (1959), Nooitgedacht (1962), Witklip (1969), Primkop (1970), Vygeboom, Da Gama (1971), Mhlume Transf (1981), Kwena (1984), Corrumana (1989), Ngodwana (1993), Driekoppies (1997), Maguga, Inyaka (2000)

Years: 1959–2012

Station	Place
X1H001	Komati River @ Hooggenoeg
X1H003	Komati River @ Tonga
X1H014	Mlumati River @ Lomati
X1H016	Buffelspruit @ Doornpoort
X1H021	Mtsoli River @ Diepgezet
X2H005	Nels River @ Boschrand
X2H006	Krokodil River @ Karino
X2H008	Queens River @ Sassenheim
X2H010	Noordkaap River @ Bellevue
X2H011	Elands River @ Geluk
X2H012	Dawsons Spruit @ Geluk
X2H013	Krokodil River @ Montrose
X2H014	Houtbosloop @ Sudwalaskraal
X2H015	Elands River @ Lindenau
X2H016	Krokodil River @ Tenbosch
X2H022	Kaap River @ Dolton
X2H024	Suidkaap River @ Glenthorpe
X2H031	Suidkaap River @ Bornmans Drift
X2H032	Krokodil River @ Weltevrede
X2H036	Komati River @ Komatipoort
X2H046	Krokodil River @ Riverside
X2H047	Swartkoppiesspruit @ Kindergoed
X3H001	Sabie River @ Sabie
X3H002	Klein Sabie River @ Sabie
X3H003	Mac-Mac River @ Geelhoutboom
X3H004	Noordsand River @ De Rust
X3H006	Sabie River @ Perry's Farm
X3H008	Sand River @ Exeter
X3H011	Marite River @ Injaka
X3H015	Sabie River @ Lower Sabie Rest Camp
X3H021	Sabie River @ Kruger Gate
E23	Incomati River @ Ressano Garcia
E43	Incomati River @ Magude

Good continuous data
Missing data/no data
Major data gaps/gauge limit

Figure 3-1. Streamflow data used on this study, with indication of time series length, data quality, missing data. Major developments in the basin, such as dams, are on the top horizontal line on the year they were commissioned; indication is made of the gauges affected by the developments by the initial letter of the dam.

Table 3-1 Hydrometric stations analysed, location, catchment area, data length and missing data

	Station	Latitude	Longitude	River and location	Catchment Area (km²)	Data Available	Period analysed for trends	Missing Data
Komati	X1H001	-26.04	31.00	Komati River @ Hooggenoeg	5499	1909 - 2012	1970-2011 (42 years)	8.0%
	X1H003	-25.68	31.78	Komati River @ Tonga	8614	1939 - 2012	1970-2011 (42 years)	6.8%
	X1H014	-25.67	31.58	Mlumati River @ Lomati	1119	1968 - 2012	1978-2011 (34 years)	0.5%
	X1H016	-25.95	30.57	Buffelspruit @ Doornpoort	581	1970 - 2012	1970-2011 (42 years)	3.4%
	X1H021	-26.01	31.08	Mtsoli River @ Diepgezet	295	1975 - 2012	1976-2011 (36 years)	2.7%
Crocodile	X2H005	-25.43	30.97	Nels River @ Boschrand	642	1929 - 2012	1970-2011 (42 years)	0.8%
	X2H006	-25.47	31.09	Krokodil River @ Karino	5097	1929 - 2012	1970-2011 (42 years)	0.1%
	X2H008	-25.79	30.92	Queens River @ Sassenheim	180	1948 - 2012	1970-2011 (42 years)	0.5%
	X2H010	-25.61	30.87	Noordkaap River @ Bellevue	126	1948 - 2012	1970-2011 (42 years)	5.7%
	X2H011	-25.65	30.28	Elands River @ Geluk	402	1956 - 1999	1957-1999 (43 years)	0.9%
	X2H012	-25.66	30.26	Dawsons Spruit @ Geluk	91	1956 - 2012	1970-2011 (42 years)	0.3%
	X2H013	-25.45	30.71	Krokodil River @ Montrose	1518	1959 - 2012	1970-2011 (42 years)	1.6%
	X2H014	-25.38	30.70	Houtbosloop @ Sudwalaskraal	250	1958 - 2012	1970-2011 (42 years)	5.1%
	X2H015	-25.49	30.70	Elands River @ Lindenau	1554	1959 - 2012	1970-2011 (42 years)	3.1%
	X2H016	-25.36	31.96	Krokodil River @ Tenbosch	10365	1960 - 2012	1970-2011 (42 years)	5.6%
	X2H022	-25.54	31.32	Kaap River @ Dolton	1639	1960 - 2012	1970-2011 (42 years)	5.7%
	X2H024	-25.71	30.84	Suidkaap River @ Glenthorpe	80	1964 - 2012	1970-2011 (42 years)	1.7%
	X2H031	-25.73	30.98	Suidkaap River @ Bornmans Drift	262	1966 - 2012	1966-2011 (46 years)	5.0%
	X2H032	-25.51	31.22	Krokodil River @ Weltevrede	5397	1968 - 2012	1970-2011 (42 years)	2.4%
	X2H036	-25.44	31.98	Komati River @ Komatipoort	21481	1982 - 2012	1983-2011 (28 years)	4.1%
	X2H046	-25.40	31.61	Krokodil River @ Riverside	8473	1985 - 2012	1986-2011 (26 years)	2.0%
	X2H047	-25.61	30.40	Swartkoppiesspruit @ Kindergoed	110	1985 - 2012	1986-2011 (26 years)	2.2%
Sabie	X3H001	-25.09	30.78	Sabie River @ Sabie	174	1948 - 2012	1970-2011 (42 years)	0.8%
	X3H002	-25.09	30.78	Klein Sabie River @ Sabie	55	1963 - 2012	1970-2011 (42 years)	0.4%
	X3H003	-24.99	30.81	Mac-Mac River @ Geelhoutboom	52	1948 - 2012	1970-2011 (42 years)	0.5%
	X3H004	-25.08	31.13	Noordsand River @ De Rust	200	1948 - 2012	1970-2011 (42 years)	3.9%
	X3H006	-25.03	31.13	Sabie River @ Perry's Farm	766	1958 - 2000	1970-1999 (30 years)	2.6%
	X3H008	-24.77	31.39	Sand River @ Exeter	1064	1967 - 2011	1968-2011 (43 years)	15.5%
	X3H011	-24.89	31.09	Marite River @ Injaka	212	1978 - 2012	1979-2011 (32 years)	7.6%
	X3H015	-25.15	31.94	Sabie River @ Lower Sabie Rest Camp	5714	1986 - 2012	1988-2011 (24 years)	8.2%
	X3H021	-24.97	31.52	Sabie River @ Kruger Gate	2407	1990 - 2012	1991-2011 (21 years)	10.8%
Lower Incomati	E23	-25.44	31.99	Incomati River @ Ressano Garcia	21200	1948 - 2011	1970-2011 (42 years)	9.0%
	E43	-25.03	32.65	Incomati River @ Magude	37500	1952 - 2011	1970-2011 (42 years)	3.5%

Table 3-2. Hydrologic parameters used in Range of Variability Approach (Richter *et al.*, 1996)

Indicators of Hydrologic Alteration Group	Regime Characteristics	Hydrological parameters
Group 1: Magnitude of monthly water conditions	Magnitude timing	Mean value for each calendar month
Group 2: Magnitude and duration of annual extreme water conditions	Magnitude duration	Annual minima and maxima based on one, three, seven, thirty and ninety day(s) mean
Group 3: Timing of annual extreme water conditions	Timing	Julian date of each annual 1-day maximum and minimum
Group 4: Frequency and duration of high/low pulses	Frequency and duration	No. of high and low pulses each year
		Mean duration of high and low pulses within each year (days)
Group 5: Rate/Frequency of water condition changes	Rates of change of frequency	Means of all positive and negative differences between consecutive daily values
		No. of rises and falls

3.3 Results

3.3.1 Rainfall

Data series for 20 rainfall stations were statistically analysed for the period 1950-2011 (Table 3-3). The variability of rainfall across the basin was confirmed to be high, both intra- and inter-annually, with a wide range between years. This variability is highest for the stations located in mountainous areas. The variability across the basin is also significant, as illustrated by the box plot of Figure 3-2.

The Spearman Trend Test revealed that only 5 of the 20 investigated stations showed significant trends of increase (2 stations) and decrease (3 stations). However, the stations that presented significant trends are also stations with lower percentage of reliability, thus it is possible that the trend identified could be affected by data infilling procedures. There was no serial correlation of annual and monthly time series. Some change points were identified using the Pettitt Test, mostly in the years 1971 and 1978 (Table 3-3). Only two stations out of the twenty studied showed significant change towards a wetter regime (Riverbank and Manhica).

Monthly rainfall does not also exhibit any clear trend at most stations. This is consistent with the larger scale analyses conducted by Schulze (2012) for South Africa and Shongwe *et al.* (2009) for Southern Africa. Mussá *et al.* (2013) studied the trends of annual and dry extreme rainfall, using the Standardized Precipitation Index (SPI) and also found no significant trends in annual rainfall extremes across the Crocodile sub-catchment.

Table 3-3. Description of rainfall stations analysed for trends, also the long term Mean Annual Precipitation (MAP) in mma^{-1}, the standard variation, and detection of trend (confidence level of 95% using Spearman Test) and occurrence change point (using Pettitt Test followed by T-test of stability of mean and F-test of stability of variance)

Analysis for the period 1950 to 2011

Name	Station ID	Latitude	Longitude	Altitude [MASL]	MAP [mm]	P Reliable [%]	Mean [mm a^{-1}]	St.Dev. [mm a^{-1}]	Trend Spearman	Pettitt
Machadodorp	0517430 W	-25.67	30.25	1563	781	79.6	773	134		
Badplaas (Pol)	0518088 W	-25.97	30.57	1165	829	90.6	817	153		
Kaapsehoop	0518455 W	-25.58	30.77	1564	1443	78.5	1461	286		Decr (1975)
Mac Mac (Bos)	0594539 W	-24.98	30.82	1295	1463	75.1	1501	287		
Spitskop (Bos)	0555579 W	-25.15	30.83	1395	1161	68.5	1197	266	Decr	Decr (1978)*
Alkmaar	0555567 W	-25.45	30.83	715	830	95.2	874	172		
Oorschot	0518859 W	-25.80	30.95	796	787	92.2	775	185		
Bosbokrand (Pol)	0595110 W	-24.83	31.07	778	982	82.4	919	297		Decr(1978)*
Pretoriuskop	0556460 W	-25.17	31.18	625	707	60.0	734	188		
Riverbank	0519310 W	-25.67	31.23	583	683	70.5	782	163	Incr	Incr (1977)**
Piggs Pig	0519448 A	-25.97	31.25	1029	1024	40.1	1075	315	Decr	Decr (1978)*
Skukuza	0596179 W	-25.00	31.58	300	560	63.1	566	140		
Riverside	0557115 W	-25.42	31.60	315	547	66.5	520	187		
Satara	0639504 W	-24.40	31.78	257	568	42.1	602	151	Incr	Incr (1971)
Fig Tree	0520589 W	-25.82	31.83	256	591	63.4	594	145	Decr	Decr (1978)*
Tsokwane	0596647 W	-24.78	31.87	262	540	66.1	544	134		Incr (1971)*
Krokodilbrug	0557712 W	-25.37	31.90	192	584	62.9	590	147		
Moamba	P821 M	-25.60	32.23	108	632	63.9	633	185		
Xinavane	P10 M	-25.07	32.87	18	853	76.2	773	241		
Manhica	P63 M	-25.40	32.80	33	883	86.2	903	275		Incr (1970)**

* Significant change with 2.5% significance level with T-Test of stability of mean

** Significant change with 2.5% significance level with T-Test of stability of mean and F-Test of stability of variance

Explanatory Note: MAP is the Mean Annual Precipitation, and P reliable is the percentage of reliable data for the rainfall station, as assessed by Lynch (2003) for the period 1905 to 1999. The mean refers to the average of total annual precipitation for the period of 1950 to 2011. On the column trend Spearman only stations that had trend significant at 95% confidence level are indicated with Decr or Incr, corresponding to decreasing or increasing trend, respectively. On the column Pettitt, the direction of change and year are indicated, as well as the significance of the change point

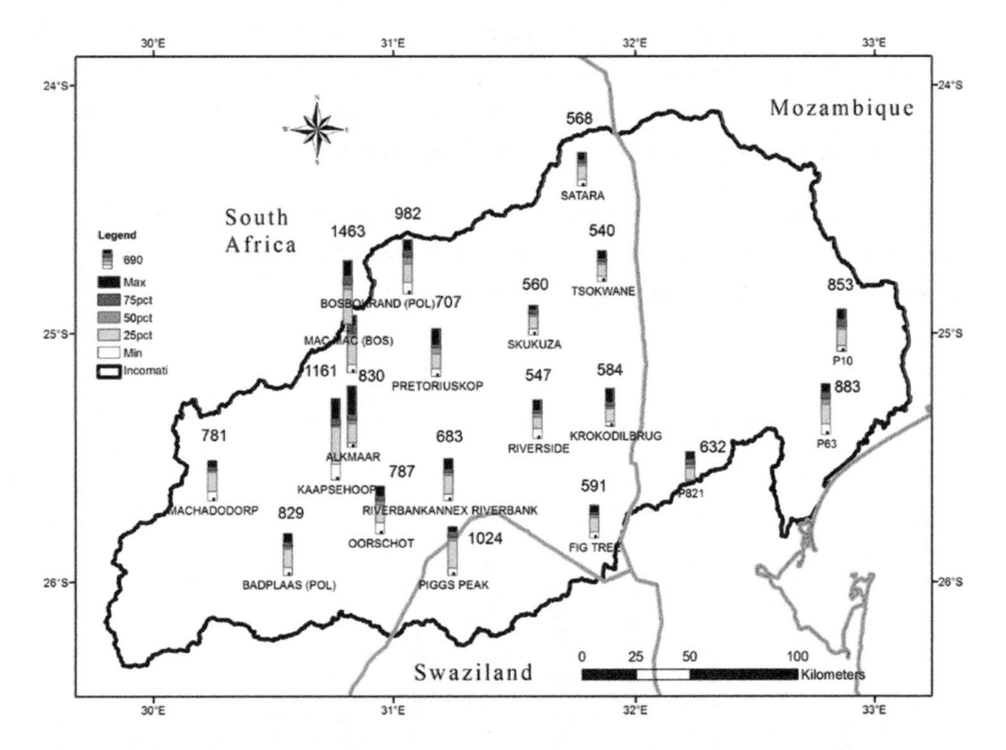

Figure 3-2 Box plot illustrating the spatial variation of annual rainfall across the Incomati Basin (median, 25%, 75% are shown by the green and red boxes; the lines illustrate the range). The stations are presented from west to east, along the basin profile.

3.3.2 Variability of streamflow

The metrics of the different hydrologic indicators were compiled as an output of the IHA analysis, which is illustrated for the gauging stations located at the outlet (or the most downstream) of each main sub-catchment in Table 3-4. The variability is described, using non-parametric statistics (median and coefficient of dispersion), because the hydrological time series are not normally distributed, but positively skewed. The coefficient of dispersion (CD) is defined as CD= (75th percentile - 25th percentile) / 50th percentile. The larger the CD, the larger the variation of the parameter will be.

The flow patterns are consistent with the summer rainfall regime, with highest flow and rainfall events associated with tropical cyclone activity in January-March.

A comparison of the flow normalized by area (Figure 3-3) for the main sub-catchments reveals that Sabie yields a higher runoff than Komati and Crocodile. This is the case because the observed streamflows include the impact of water abstractions and streamflow reduction activities, which are more intense in the Komati and Crocodile sub-catchments (Hughes and Mallory, 2008; Mallory and Hughes, 2012).

Another aspect to note is that the flows of February are likely to be higher than observed records, but high streamflow extremes are not fully captured by the current monitoring network, due to gauging stations limitations.

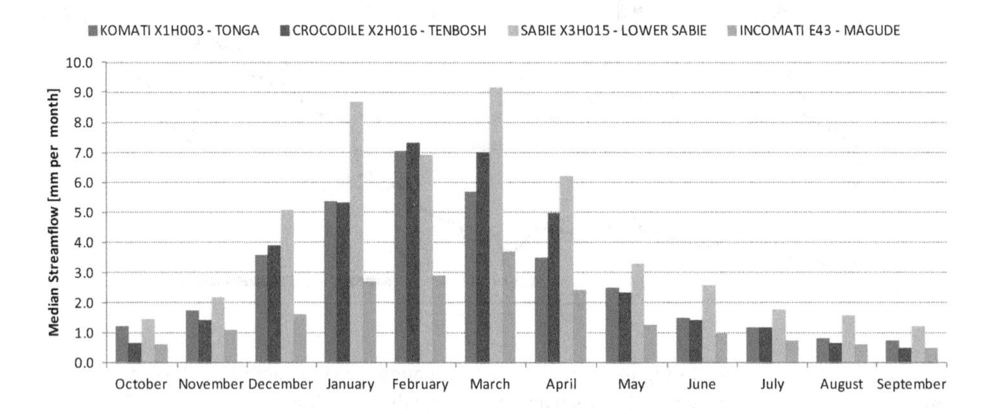

Figure 3-3. Median of observed daily streamflow for the gauges located at the outlet of major sub-catchments Komati, Crocodile, Lower Sabie and Incomati (based on daily time series from 1970 to 2011)

Table 3-4. Hydrological indicators of main sub-catchments

STREAMFLOW INDICATORS	UNITS	KOMATI X1H003 - TONGA		CROCODILE X2H016 - TENBOSH		INCOMATI X2H036 - KOMATIPOORT		SABIE X3H015 - LOWER SABIE		INCOMATI E43 - MAGUDE	
Period of Analysis:		1970-2011 (42 years)		1970-2011 (42 years)		1983-2011 (28 years)		1988-2011 (24 years)		1970-2011 (42 years)	
Drainage area	km^2	8614		10365		21481		5714		37500	
		Median	CD	Median	CD	Median	CD	Median	CD	Median	CD
Annual*	m^3s^{-1}	**16.94**	2.14	**21.35**	1.97	**34.28**	2.11	**17.35**	2.31	**47.44**	2.01
October	m^3s^{-1}	3.95	1.47	2.54	1.88	2.24	1.87	3.08	0.92	8.72	1.21
November	m^3s^{-1}	5.72	1.94	5.75	2.35	7.09	3.88	4.81	1.09	16.14	1.49
December	m^3s^{-1}	11.46	2.09	15.07	1.48	18.79	2.63	10.83	1.49	22.91	2.90
January	m^3s^{-1}	17.26	1.82	20.68	1.47	34.47	1.52	18.52	1.35	37.96	1.35
February	m^3s^{-1}	25.09	1.95	31.37	2.01	29.77	2.80	16.33	1.84	45.09	3.21
March	m^3s^{-1}	18.33	1.74	27.15	1.63	42.15	1.90	19.51	2.30	51.75	2.32
April	m^3s^{-1}	11.64	1.74	19.82	1.37	24.10	2.13	13.69	1.13	34.90	2.03
May	m^3s^{-1}	8.03	1.41	9.11	1.68	9.98	2.16	7.04	1.64	17.85	1.86
June	m^3s^{-1}	4.96	1.90	5.66	1.62	7.10	2.45	5.64	1.25	14.04	1.44
July	m^3s^{-1}	3.77	1.98	4.56	1.48	4.72	2.28	3.79	1.18	10.41	1.47
August	m^3s^{-1}	2.67	1.63	2.63	1.71	2.51	1.35	3.40	1.08	8.46	1.41
September	m^3s^{-1}	2.43	1.47	2.08	1.81	2.24	1.51	2.69	1.15	7.06	1.11
1-day minimum	m^3s^{-1}	0.31	4.04	0.24	2.64	0.14	5.29	1.45	1.13	2.49	1.48
3-day minimum	m^3s^{-1}	0.38	3.38	0.32	2.16	0.25	3.76	1.53	1.08	2.71	1.76
7-day minimum	m^3s^{-1}	0.59	2.55	0.40	2.88	0.33	4.35	1.60	1.16	3.01	1.61
30-day minimum	m^3s^{-1}	1.46	2.13	1.52	1.79	1.29	2.08	2.01	1.12	4.84	1.37
90-day minimum	m^3s^{-1}	3.69	1.47	3.45	1.34	3.17	2.09	3.02	1.23	8.14	1.38
1-day maximum	m^3s^{-1}	134.4	1.26	142.2	1.38	274.3	1.00	113	2.51	381.5	1.80
3-day maximum	m^3s^{-1}	102.9	1.50	126.9	1.33	232.9	1.15	87.62	2.60	344.1	1.74
7-day maximum	m^3s^{-1}	81.79	1.59	107.4	1.20	201.4	1.13	62.55	2.27	273.7	1.56
30-day maximum	m^3s^{-1}	54.39	1.45	76.98	1.28	109.6	1.33	37.66	1.93	156.7	1.45
90-day maximum	m^3s^{-1}	39.19	1.33	45.08	1.16	68.69	1.71	28.06	1.47	102	1.32
Date of minimum	Julian Date	275	0.10	274	0.12	281.5	0.15	278.5	0.06	290.5	0.21
Date of maximum	Julian Date	38.5	0.16	33	0.11	35.5	0.19	20.5	0.17	39.5	0.14
Low pulse count	No	6	1.63	4	1.63	5	1.55	4	1.00	3	1.33
Low pulse duration	Days	5.5	1.41	5	1.60	3.5	0.71	6.5	1.69	6.75	2.09
High pulse count	No	6	0.75	4	1.25	5	0.95	4	0.69	4	0.75
High pulse duration	Days	4	1.31	4	2.13	4.5	1.28	5	2.10	8.5	1.03
Rise rate	m^3s^{-1}	0.7095	1.39	0.64	0.98	1.161	1.38	0.404	1.12	1.058	1.43
Fall rate	m^3s^{-1}	-0.7295	-0.98	-0.61	-0.78	-1.38	-1.28	-0.2398	-1.10	-0.6278	-2.31
Number of reversals	No	111.5	0.26	113	0.42	121	0.18	95	0.29	86	0.49

* On the annual statistics mean and coefficient of variation were used

CD is the coefficient of dispersion

Table 3-5. Trends of the hydrological indicators for the period 1970-2011. In bold are significant trends at 95% confidence level.

STREAMFLOW INDICATORS	KOMATI X1H003 - TONGA		CROCODILE X2H016 - TENBOSH		SABIE X3H015 - LOWER SABIE		INCOMATI X2H036 - KOMATIPOORT		INCOMATI E43 - MAGUDE	
Period of Analysis:	1970-2011 (42 years)		1970-2011 (42 years)		1988-2011 (24 years)		1983-2011 (28 years)		1970-2011 (42 years)	
Drainage area [km²]	8614		10365		5714		21481		37500	
	Slope	*Pvalue*	Slope	*Pvalue*	Slope	*Pvalue*	Slope	*Pvalue*	Slope	*Pvalue*
October	-0.285	*0.05*	-0.052	*0.5*	0.017	*0.5*	-0.017	*0.5*	-0.313	*0.25*
November	-0.254	*0.1*	-0.006	*0.5*	0.263	*0.5*	0.020	*0.5*	-0.165	*0.5*
December	-0.194	*0.5*	-0.090	*0.5*	0.199	*0.5*	0.783	*0.5*	-0.087	*0.5*
January	-0.437	*0.5*	-0.023	*0.5*	1.493	*0.25*	1.979	*0.25*	-0.960	*0.5*
February	-1.027	*0.1*	-0.927	*0.25*	0.544	*0.5*	-0.486	*0.5*	**-2.847**	*0.05*
March	-0.360	*0.5*	-0.397	*0.5*	0.390	*0.5*	-0.112	*0.5*	-1.346	*0.5*
April	-0.082	*0.5*	-0.007	*0.5*	0.899	*0.25*	1.532	*0.25*	-0.195	*0.5*
May	-0.225	*0.1*	-0.045	*0.5*	0.416	*0.5*	0.788	*0.5*	-0.365	*0.5*
June	**-0.215**	*0.025*	0.059	*0.5*	0.270	*0.5*	0.470	*0.5*	-0.045	*0.5*
July	**-0.179**	*0.005*	0.060	*0.5*	0.219	*0.5*	0.171	*0.5*	-0.039	*0.5*
August	-0.074	*0.1*	0.105	*0.5*	0.134	*0.25*	0.312	*0.5*	0.090	*0.5*
September	-0.029	*0.5*	0.134	*0.5*	0.081	*0.5*	0.218	*0.5*	0.166	*0.25*
1-day minimum	**-0.027**	*0.025*	-0.015	*0.25*	0.061	*0.1*	0.003	*0.5*	**0.139**	*0.001*
3-day minimum	**-0.029**	*0.025*	-0.015	*0.25*	0.061	*0.1*	0.004	*0.5*	**0.127**	*0.005*
7-day minimum	**-0.038**	*0.05*	-0.015	*0.5*	0.064	*0.1*	0.004	*0.5*	0.094	*0.05*
30-day minimum	**-0.069**	*0.025*	-0.025	*0.25*	0.058	*0.25*	0.033	*0.5*	0.054	*0.5*
90-day minimum	**-0.115**	*0.01*	-0.059	*0.25*	0.131	*0.1*	0.038	*0.5*	-0.054	*0.5*
1-day maximum	-5.143	*0.25*	-5.425	*0.25*	-2.743	*0.5*	-12.070	*0.25*	**-10.580**	*0.025*
3-day maximum	-3.749	*0.25*	-3.670	*0.25*	-1.379	*0.5*	-8.171	*0.5*	**-9.254**	*0.025*
7-day maximum	-2.361	*0.25*	-2.427	*0.25*	0.014	*0.5*	-3.742	*0.5*	**-6.722**	*0.05*
30-day maximum	-1.022	*0.25*	-1.023	*0.25*	0.662	*0.5*	0.092	*0.5*	-3.400	*0.1*
90-day maximum	-0.671	*0.25*	-0.576	*0.5*	0.789	*0.5*	0.934	*0.5*	-2.147	*0.25*
Number of zero days	0.690	*0.25*	-0.005	*0.5*	0	*0.5*	0.032	*0.5*	-0.080	*0.5*
Base flow index	-0.001	*0.25*	0.000	*0.5*	0.004	*0.5*	0.001	*0.25*	**0.007**	*0.001*
Date of minimum	-0.686	*0.5*	0.354	*0.5*	0.548	*0.5*	-0.420	*0.5*	1.374	*0.5*
Date of maximum	0.817	*0.5*	0.347	*0.5*	-3.222	*0.5*	0.288	*0.5*	0.617	*0.5*
Low pulse count	0.132	*0.1*	**0.238**	*0.001*	-0.045	*0.5*	0.185	*0.5*	0.043	*0.25*
Low pulse duration	0.068	*0.5*	-0.140	*0.5*	-0.669	*0.1*	-0.297	*0.25*	-0.602	*0.5*
High pulse count	**-0.127**	*0.005*	0.007	*0.5*	-0.023	*0.5*	-0.096	*0.25*	-0.068	*0.05*
High pulse duration	0.029	*0.5*	**-1.263**	*0.01*	1.081	*0.25*	0.144	*0.5*	-0.103	*0.5*
Rise rate	-0.007	*0.5*	-0.008	*0.5*	0.005	*0.5*	0.017	*0.5*	**-0.034**	*0.05*
Fall rate	0.003	*0.5*	**-0.013**	*0.05*	-0.007	*0.5*	-0.012	*0.5*	-0.007	*0.5*
Number of reversals	0.574	*0.1*	**1.083**	*0.01*	0.723	*0.5*	0.560	*0.5*	**0.764**	*0.005*

3.3.3 Trends in streamflow

Figure 3-4 presents a spatial plot of trends for selected hydrological indicators for the periods 1970-2011 (Figure 3-4a) and 1950-2011 (Figure 3-4b). The significant trends are highlighted with a circle. Table 3-5 presents the slope of the trend lines and P values for the gauges located at the outlet, or the most downstream point of each main sub-catchment. There is a significant trend of decreasing mean flow in October at almost all stations, especially the ones located on the main stem of the Crocodile and the Komati Rivers (Figure 3-6). October is the month of the start of the rainy season, when the dam levels are lowest and irrigation water requirements highest (DWAF, 2009c; ICMA, 2010).

This trend is consistent with the decreasing trends of minimum flows, as exemplified by the 7-day minimum. In contrast, it can be seen that the count of low pulses increased significantly in many gauges, which indicates the more frequent occurrence of low flows. Another striking trend is the significant increase of the number of reversals at almost all stations. Reversals are calculated by dividing the hydrologic record into "rising" and "falling" periods, which correspond to periods in which daily changes in flows are either positive or negative, respectively. The number of reversals is the number of times that flow switches from one type of period to another. The observed increased number of reversals is likely due to the effect of flow regulation and water abstractions.

The significant trends (95% confidence level) of the various indicators were counted per station and plotted on a map (Figure 3-5). Most significant decreasing trends occur in the Komati and Crocodile systems, which are also the most stressed sub-catchments. An interesting aspect is that some of the trends cross-compensate each other. Some of the positive trends occurring on the tributaries of the Crocodile, for example, the October Median Flow and baseflow are cancelled out when moving down the main stem of the river.

The cross-compensation can also be observed at basin-scale on the Sabie, where the trends of decreasing flows are not so frequent or significant. It is likely that this occurs because the majority of the Sabie falls under the conservation area of the Kruger National Park (KNP) and therefore fewer abstractions occur compared to other sub-catchments, as illustrated in Table 2-2. The KNP has been playing an important role in the catchment management fora set up by the Inkomati Catchment Management Agency (ICMA), which concern the provision of environmental minimum flows, in order to maintain ecosystem services and biodiversity in the Park (Pollard *et al.*, 2012; Riddell *et al.*, 2014b).

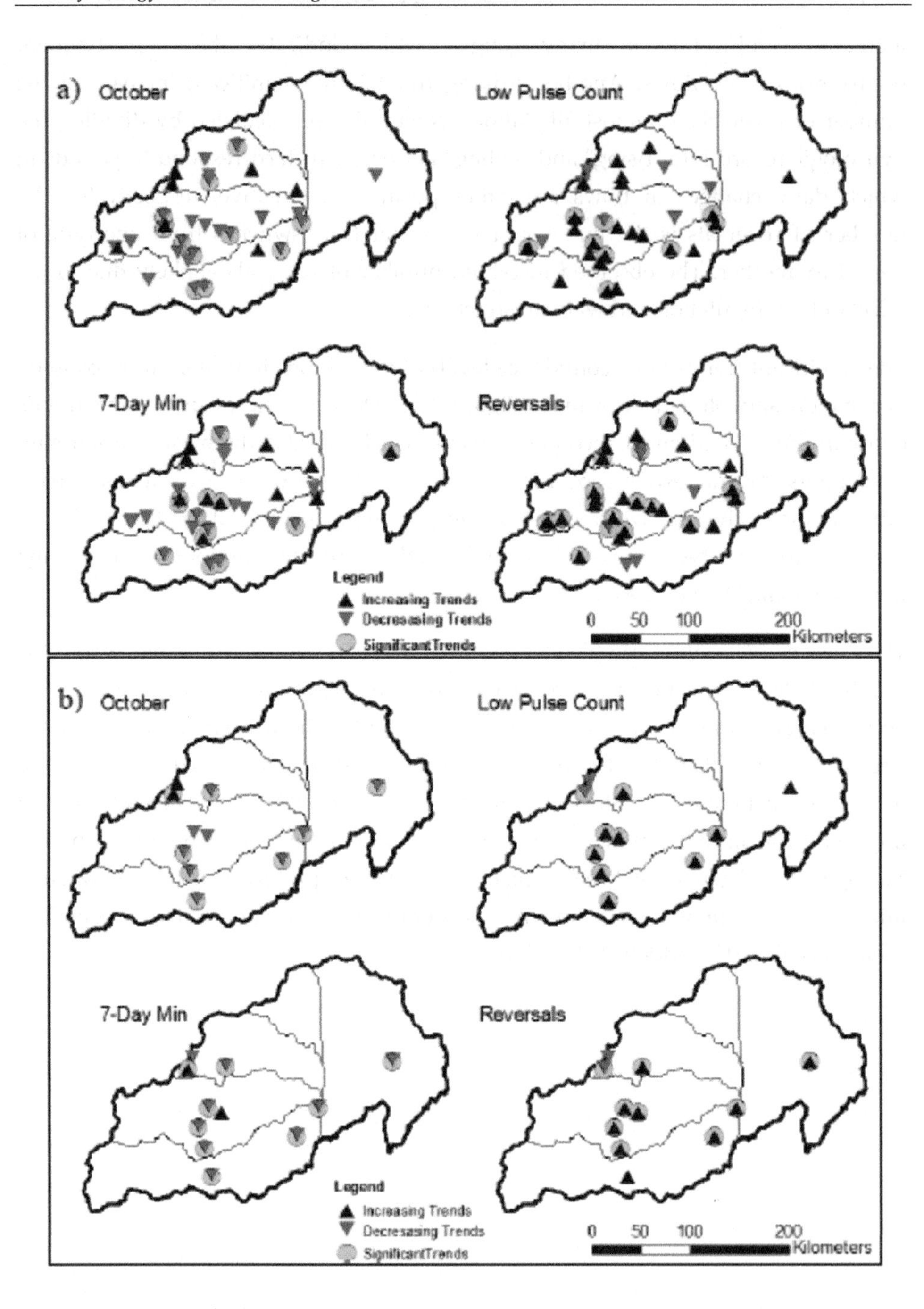

Figure 3-4. Trends of different indicators of streamflow: a) for period 1970-2011; b) for period 1950-2011

Table 3-5 illustrates that many of the trends observed in the Sabie sub-catchment contrast those observed in the Komati and Crocodile sub-catchments. Thus, the trends observed in Magude (station E43) in Mozambique are the result of a combination of the positive effect of the conservation approach of KNP on the Sabie, and the negative effect of flow reductions in the Crocodile and the Komati.

The Komati sub-catchment (at Tonga gauging station, X1H003) is where most negative trends occur, particularly significant during the months of October, June and July (Table 3-5 and Figure 3-5). At the downstream end of the Crocodile (at Tenbosch gauge, X2H016) similar trends are not visible, because of cross-compensations: the Kaap and Elands tributaries both have significant decreasing trends of their mean monthly flows, as well as the low flows; the Kwena Dam, located on the main stem of the Crocodile, on the other hand, is managed in a way to augment the flows during the dry season.

It is important to note that these trends are even more pronounced, when longer time series are considered. Two examples from the Crocodile Basin are presented below.

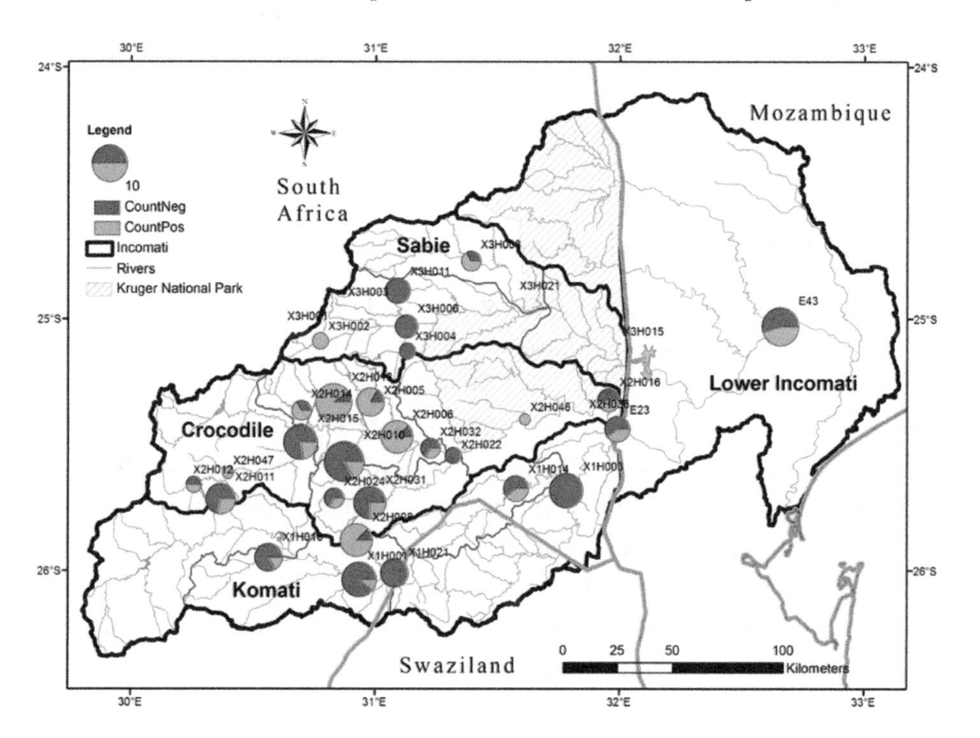

Figure 3-5. Count of significant trends. Declining trends are in red and increasing trends in green. The size of the pie is proportional to the total number of significant trends.

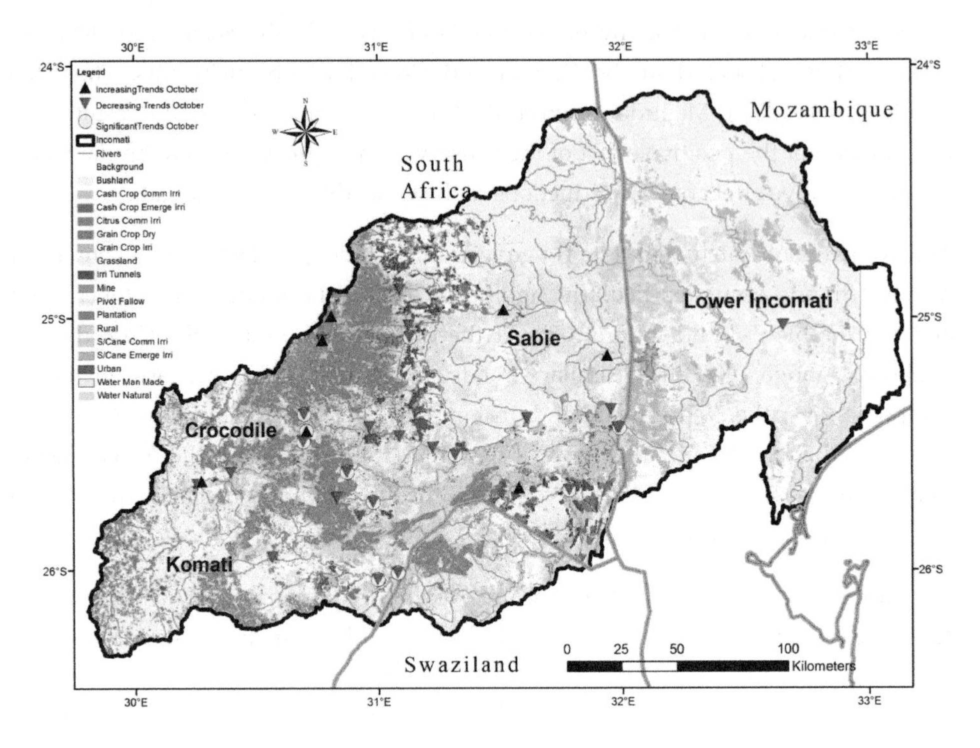

Figure 3-6. Land use land cover map of Incomati (ICMA, 2010; TPTC, 2010) and streamflow trends in the month of October

3.3.3.1 Example of decreasing trends: Noord Kaap X2H010

The Noord Kaap gauging station (X2H010), located on a tributary of the Crocodile sub-catchment, displays the most intriguing trends. Out of the 33 IHA indicators, this gauge had 12 significant trends, 10 of which negative, indicating a major shift in flow regime. The decreasing trends occur in all months, but are more pronounced during low flow months, particularly September (Figure 3-7) and October. There is a significant decrease of high flows and small floods and an increase of extreme low flows. However, there is no record of the presence of a dam or major infrastructure (DWAF, 2009b). The areal rainfall for the drainage area of this station did not show a significant decreasing trend, which suggests that the reduction observed in streamflow should be a result of land use change, namely, conversion to forestry and irrigated land. Figure 3-8 illustrates the comparison of median monthly flows for the two periods. From the analysis of land use changes over time (Table 2-4), the sharp decrease of mean monthly flows during the 1960s coincides with an increase of the area under irrigated agriculture. During the 1970s there was also a great increase of area under forestry, namely, Eucalyptus (DWAF, 2009c). Commercial forestry consumes more water through evaporation than the native vegetation it replaces. Therefore, under the South African National Water Act, a commercial forest plantation is considered a Streamflow Reduction Activity (SFRA) and must be licensed as a water user (Jewitt, 2002; Jewitt, 2006b). A recent study by van Eekelen *et al.* (2015) finds that stream flow reduction due to forest plantations may be twice or even three times more than that allowed by the Interim IncoMaputo Agreement.

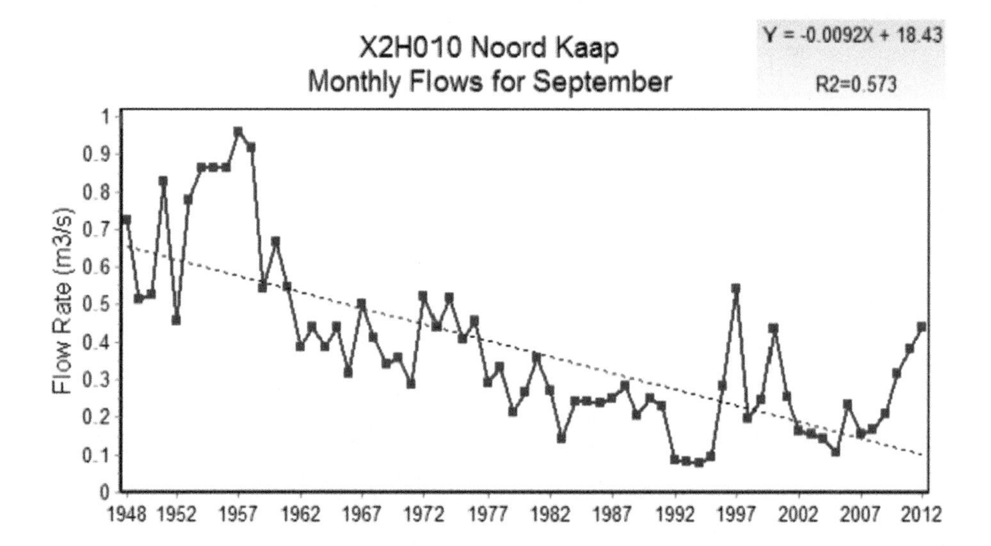

Figure 3-7. Plot of median monthly flows for September for the entire time series (1949-2011) on the Noord Kaap Gauge, located on the Crocodile sub-catchment.

3.3.3.2 Impact of the Kwena Dam on streamflows of the Crocodile River

The Kwena Dam, commissioned in 1984, is the main reservoir on the Crocodile system, located in the upper part of the catchment,. The dam is used to improve the assurance of supply of water for irrigation purposes in the catchment. The Montrose station (X2H013) is located 35 kilometres downstream of this dam. The two-period (1959-1984 and 1986-2011) analysis illustrates the main impacts of Kwena Dam on the river flow regime, namely the dampening of peak flows and an increase of low flows (Figure 3-9). These results are consistent with the analysis conducted by Riddell *et al.* (2014b), which found significant alterations of natural flow regime in the Crocodile Basin over the past 40 years. Similar impacts were found in studies in different parts of the world (Richter *et al.*, 1998; Bunn and Arthington, 2002; Maingi and Marsh, 2002; Birkel *et al.*, 2014a).

It can be seen that the Kwena Dam is managed to augment the low flows and attenuate floods. This change in the flow regime influences the streamflow along the main stem of the Crocodile River, but as tributaries join and water is abstracted, the effect is reduced. At the outlet at Tenbosch station X2H016 (Figure 3-5 and Table 3-5), the effects of flow regulation and water abstractions have counter-balanced the contrasting trends observed upstream.

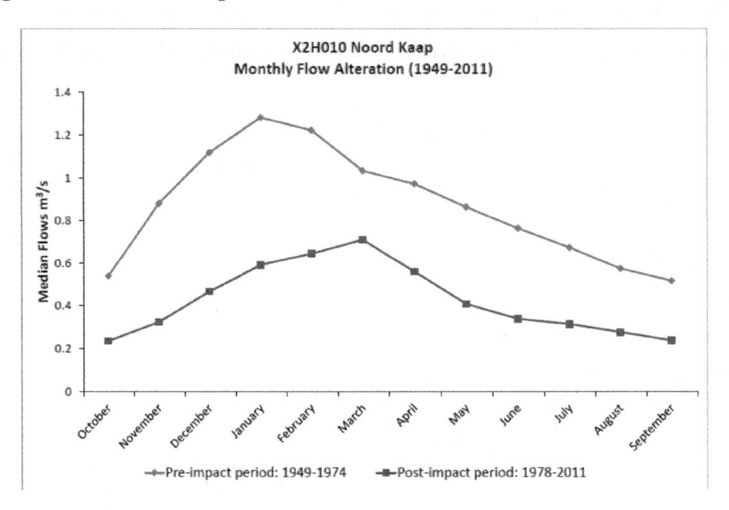

Figure 3-8. Plot of median monthly flows for 2 periods (1949- 1974 and 1978-2011) on the Noord Kaap Gauge, located on the Crocodile sub-catchment.

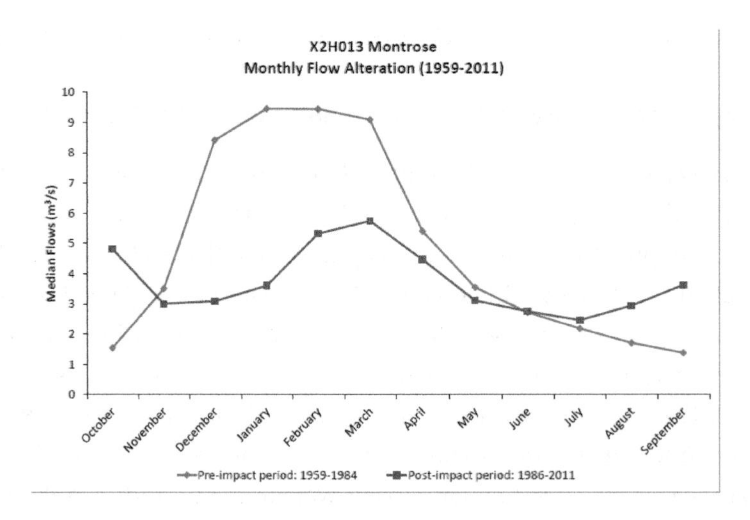

Figure 3-9. Impact of Kwena Dam (commissioned in 1984) on streamflows of the Crocodile River, Montrose Gauge X2H013

3.3.3.3 Impact of anthropogenic actions

As can be seen from water use information, the impacts of land use change and water abstractions are the main drivers of changes in the flow regime of the Incomati River. However, the situation is variable along the catchment. In the Sabie system, in spite of large areas of commercial forestry in the headwaters, the mean, annual and low flows do not show significant trends (Table 3-5). This can be explained by the fact that most of the forestry area was already established during the period of analysis (1970-2011)(DWAF, 2009d). The fact that a large proportion of the Sabie sub-catchment is under conservation land use (KNP and other game reserves) also plays an important role in maintaining the natural flow regime.

On the Crocodile, however, irrigated agriculture, forestry and urbanization were the most important anthropogenic drivers. They affect the streamflow regime, the water quantity and possibly the water quality as well (beyond the scope of this analysis). This has important implications when environmental flow requirements and minimum cross-border flows need to be adhered to. Pollard and du Toit (2011b) and Riddell *et al.* (2014b) have demonstrated that the Crocodile River is not complying with the environmental flow requirements during most of the dry season at the outlet.

On the Komati, the strategic water uses, which have first priority (such as the water transfers to ESKOM plants in the Olifants Catchment and to irrigation schemes in the Umbeluzi) (Nkomo and van der Zaag, 2004; DWAF, 2009e), have a high impact on streamflows. Because of other water allocations, for irrigation, forestry and other industries, steady trends of decreasing flows could be identified. This is another system where the environmental flows and cross-border requirements are often not met during the dry season (Pollard and du Toit, 2011b; Mukororira, 2012; Riddell *et al.*, 2014b).

3.4 Discussion

3.4.1 Limitations of this study

The available data series have some gaps, especially during high flow periods. Because of this, the analysis of high flow extremes is highly uncertain. For the trend analysis, the period of common data followed the construction of several impoundments and other developments.

Another challenge is the disparity of data availability across the different riparian countries. In Mozambique, only two gauges had reliable flow data for this analysis, representing the entire Lower Incomati system. The rivers Massintoto, Uanetse and Mazimechopes, in Mozambique do not have active flow gauges. There is definitely a need to strengthen the hydrometric monitoring network in the Mozambican part of the basin, as well as on the tributaries originating in the Kruger National Park.

3.4.2 What are the most striking trends and where do they occur?

The analysis resulted in the identification of major trends, including:

- Decreasing trends of the magnitude of monthly flow (significant for low flow months, e.g. October), minimum flow (1-, 3-, 7-, 30 and 90-day minimum) and the occurrence of high flow pulses;
- Significant increasing trends of the magnitude of monthly flow (August and September) in some locations in the Crocodile and Sabie, and on the occurrence of flow reversals basin wide;
- Some gauges showed no significant change or no clear pattern of change on the parameters analysed. These are mainly gauges located on the Sabie, which by 1970 had already established the current land use.

In the Komati system, the flow regulation and water abstractions have strong impacts on streamflow. Most gauges are severely impacted and it is quite difficult to characterize natural flow conditions. Flow regulation has the largest impact on low flow and minimum flows. In the Komati, irrigated agriculture is significant, particularly sugar-cane. The upstream dams of Nooitgedacht and Vygeboom are mainly used to supply cooling water to ESKOM power stations outside the basin; thus this water is exported and not used within the basin.

In the Crocodile system, flow regulation by the Kwena Dam has attenuated extreme flow events. The high flows are reduced and the low flows generally increase, leading to reverse seasonality downstream. Reverse seasonality is the change in timing of hydrograph characteristics, for example the occurrence of low flows in the wet season or high flows in the dry season. The Kwena Dam is used to improve the assurance of the supply of water for irrigation purposes in the catchment. However, Noord Kaap station X2H010, on the headwater tributary of the Crocodile, experiences a significant reduction of flows, in the monthly flow, the flow duration curves and the low flow parameters. These changes were compared with the increase in the area under forestry in the sub-catchment, as well as with the increase in irrigation. The comparison revealed that the land use change was the main driver of the flow alteration.

In the Sabie system, most gauges did not show significant trends. This is most likely due to fewer disturbances compared to the other catchments, lower water demands, few water abstractions and large areas under conservation

3.4.3 Implications of this findings for water resources management

The results of this study illustrate some hotspots where more attention should be put in order to ensure provision of water to society and the environment. When the analysis of trends is combined with that of land use of the basin (Figure 3-6), it is clear that the majority of gauges with decreasing trends are located in areas were forestry or irrigated agriculture dominate the land use and where conservation approaches are less prevalent. The presence of water management infrastructure (dams) highly influence the flow regime.

For the management of water resources in the basin, it is important to note some clear patterns, illustrated by the Sabie, Crocodile and Komati. The Sabie flows generated in the upper parts of the catchment are largely unaltered until the outlet, whilst in other rivers flows are highly modified. This suggests that the use of the conservation approach through the Strategic Adaptive Management of the Kruger National Park (KNP) and Inkomati Catchment Management Agency (ICMA), which are stronger on the Sabie, can be very beneficial to keep environmental flows in the system. It is important to consider not only the magnitude of flows, but their duration and timing as well.

Dams provide storage, generate hydropower and attenuate floods in the basin, but have impacts downstream, such as the change of mean monthly flows, the reversal of seasonality and the trapping of sediments, which can all hamper the health of downstream ecosystems. The recently concluded Mbombela Reconciliation Strategy (Beumer and Mallory, 2014) strongly recommends the construction of new dams in South Africa, including one at Mountain View in the Kaap sub-catchment. The plans for these developments were made when Swaziland was not yet fully utilizing its allocation under the Piggs Peak Agreement and Interim IncoMaputo Agreement (TPTC, 2010). Experiences of other countries around the world show that dam construction has many, often wide-ranging and long-term social and ecological impacts that often are negative and that frequently are irreversible, including the social upheaval caused by the resettlement of communities, loss of ecosystems and biodiversity, increased sediment trapping, irreversible alteration of flow regimes and the prohibitive cost of decommissioning (see for an overview (Tullos *et al.*, 2009; Moore *et al.*, 2010)). It is therefore important to fully explore alternative options before deciding of the construction of more large dams. So alternative possibilities of restoring natural stream flows and/or increasing water storage capacity should be further investigated and adopted. These alternatives could include aquifer storage, artificial recharge, rainfall harvesting, decentralized storage, and reducing the water use of existing uses and users, including irrigation, industry and forest plantations. The operation rules of existing and future dams should also include objectives to better mimic crucial aspects of the system's natural variability.

Given the likely increase of water demands due to urbanization and industrial development, it is also important that water demand management and water conservation measures are implemented in the basin. For example, there could be systems to reward users that use technology to improve their water use efficiency and to municipalities that encourage their users to lower water use.

This study also shows the complexity of water resource availability and variability. The complexity is even more relevant, considering that this is a transboundary basin with international agreements regarding minimum cross-border flows and maximum development levels that have to be adhered to (Nkomo and van der Zaag, 2004; Pollard and du Toit, 2011a; Riddell *et al.*, 2014b).

There is a great discrepancy of data availability between different riparian countries. It is important that Mozambique, in particular, improves its monitoring network, in order to better assess the impact of various management activities occurring upstream on the state of water resources. Monitoring of hydrological extremes should receive more attention, with focus on increasing the accuracy of recording flood events. The improvement of the monitoring network can be achieved by various means, such as:

- Water management institutions collaborate more intensely with academic and consultant institutions;
- Develop realistic plans to improve monitoring and data management;
- Learn from other countries/institutions that have adequate monitoring in place;
- Use modern ICT and other technologies, which may become cheaper and more accessible;

- Involve more stakeholders and citizens in data collection.

3.5 Conclusions

The research conducted reveals the dynamics of streamflow and their drivers in a river basin.

The statistical analysis of rainfall data revealed no consistent significant trend of increase or decrease for the studied period. The analysis of streamflow, on the other end, revealed significant decreasing trends of some streamflow indicators, particularly the median monthly flows of September and October, and low flow indicators. This study concludes that land use and flow regulation are the largest drivers of streamflow trends in the basin. Indeed, over the past 40 years the areas under commercial forestry and irrigated agriculture have increased over four times, increasing the consumptive water use, basin wide.

The study recommends that strategic adaptive management adopted by the Kruger National Park and Inkomati Catchment Management Agency, should be further employed in the basin. Water demand management and water conservation should be alternative options to the development of dams, and should be further investigated and established in the basin. Land use practices, particularly forestry and agriculture, have a significant impact on water quantity of the basin; therefore, stakeholders from these sectors should work closely with the water management institutions, when planning for future developments and water allocation plans.

Considering the high spatial variability of the observed changes, no unified approach will work, but specific tailor-made interventions are needed for the most affected sub-catchments and main catchments. Future investigations should conduct a careful basin-wide assessment of benefits derived from water use, and assess the first priority water uses, in particular commercial forest plantations, which *de facto*, not *de jure*, are priority users.

[illegible] and drivers of this variation.

Conclusions

[illegible]

The statistical analysis of rainfall data [illegible]

[illegible]

[illegible]

4

Isotopic and Hydrochemical River Profile of Incomati River Basin

In this chapter, water quality of the Incomati basin is described, from snapshot sampling and secondary data of the Department of Water and Sanitation, Water Management System. Water quality parameters and tracers were used to improve the understanding of hydrological processes on the Incomati River basin. Given their physical properties, the use of tracers particularly environmental isotopes Deuterium and Oxygen 18 is an innovative way of studying hydrological processes. Spatial snapshot sampling was used to give an instantaneous picture of the catchments' hydrochemistry and isotopes over two years, during wet and dry seasons. Furthermore, historical water quality data were explored to understand trends over time of water quality parameters. Results revealed increase of some of the analysed parameters from upstream to downstream of the river profiles.

This chapter is partly based on: A.M.L. Saraiva Okello, E. Riddell, S. Uhlenbrook, I. Masih, G. Jewitt, P. van der Zaag, S. Lorentz, 2012. Isotopic and Hydrochemical River Profile of the Incomati River Basin. Conference proceedings 13th WaterNet/WARFSA/GWP-SA Symposium, Johannesburg, South Africa

4.1 Introduction

The Incomati River basin has experienced several issues related with water quality, particularly due to the influence of land use changes, increased population, irrigation return flows, mining activities and flow regulation through dams and weirs. Deksissa *et al.* (2003) reported eutrophication and salinity as critical issues in the Crocodile catchment, as a result of farming, mining, industries and urbanization. They further explain that flow regulation affects mostly low flows in the catchment, by reducing the dilution capacity of streams. Mhlanga et al. (2006) and Lorentzen (2009) discuss the impacts of sugarcane farming in the Incomati basin, particularly in the Komati and Lower Crocodile catchments. LeMarie *et al.* (2006) and Macamo *et al.* (2015) described the reduction of mangrove area and the negative impact on Incomati estuary ecosystems, attributed to human activities (deforestation, altered flow regime). According to Hoguane and Antonio (2016), the minimum cross border flow established in the Piggs Peak agreement (2 m^3/s) is not sufficient to meet the minimum environmental flow requirements of the estuary. They estimated using hydrodynamic and water quality models that at least 20 m^3/s would be required to prevent salinity intrusion.

Water quality data can be used to bridge gaps in understanding of hydrological processes. Tracer methods provide excellent tools for examining hydrological processes, particularly runoff generation processes. The use of natural tracers such as isotopes or geochemical tracers provides valuable information about runoff components and their formation and dynamics at the catchment scale (Uhlenbrook and Leibundgut, 2000; Soulsby *et al.*, 2004; Tetzlaff and Soulsby, 2008; Capell *et al.*, 2011). This means that a whole hydrological system of catchment can be investigated. Environmental tracers are useful because of their conservative behaviour. Stable water isotopes are natural tracers of water movement. Isotopes ratios of hydrogen (2H) and oxygen (18O) within water molecules themselves are particularly useful, as comparison of stream waters with precipitation can indicate the nature and timing of catchment flow paths (Tetzlaff *et al.*, 2007; Wissmeier and Uhlenbrook, 2007). They can provide useful information to quantify and understand the partition of evaporation, transpiration and soil water (Yepez *et al.*, 2003). Diamond and Jack (2018) used stable isotopes to quantify evaporation, water abstractions and tributary contribution in the neighbouring Gariep River basin. They performed snapshot sampling from source to sea during the dominant flow regime (regular to low flow) across 2000 km. They then applied the principles of fractionation, mass balance and water balance to quantify evaporation from rivers and reservoirs, tributary contribution and water abstractions along the river. Abiye *et al.* (2013) provides a useful summary of isotope applications in Southern Africa over the past decades. They report case studies of isotopes used to quantify recharge, leakage from dams, evaporation, source delineation, catchment hydrology, non-point source pollution, among others, in several locations in Southern Africa.

This research aims at providing baseline profile of isotopes and hydrochemistry of the Incomati basin, from snapshot sampling and historical data analysis, to inform hydrological process understanding in the basin.

4.2 Methods and data

4.2.1 Study Area

The Incomati River basin's location, physiographic and socio-economic characteristics are presented on Chapter 2.

4.2.2 Spatial snapshot and parameters analysed

Spatial snapshot sampling was conducted on the Incomati River basin at the end of the wet season (February 2011 and March 2012) and during stable low flows (July 2011 and 2012). From March 2012 to April 2013 snapshot sampling of key locations was conducted monthly in conjunction with ICMA and KNP, as part of their monthly monitoring programs. From November 2013, an automatic water sampler was installed at the outlet of Kaap catchment, to sample water at finer temporal scale and capture events (Chapter 6). The aim was to get a hydrochemical and isotopic mapping of the catchment, in order to identify sources, pathways and response times of components of discharge contributing to upstream water flow. The parameters analysed include:

- Electrical conductivity (EC), Temperature, pH, Oxidation Reduction Potential (ORP)- measured in situ with a Multi-sensor;
- Alkalinity measured in situ by Gran titration method;
- Cations: Ca^{2+}, Mg^{2+}, Na^{+}, K^{+} - measured in the Chemistry Laboratory, University of KwaZulu-Natal (UKZN) and at IHE Delft, the Netherlands;

- Anions: HCO_3^-, NO_3^-, SO_4^{2-}, - measured in the Chemistry Laboratory at UKZN and at IHE Delft;
- Oxygen 18 and Deuterium - measured in an isotope laboratory at UKZN and at IHE Delft.

Sampling locations (Figure 4-1) where chosen from the water quality database of the Department of Water and Sanitation (DWS) in South Africa, and based on the water quality study conducted by JIBS (2001). They represent locations close to dams, DWS weirs, and main river confluences. Geology and land use were also considered. Where there was a significant change of geology and/or land uses, a greater number of samples was collected to capture the influence of these changes.

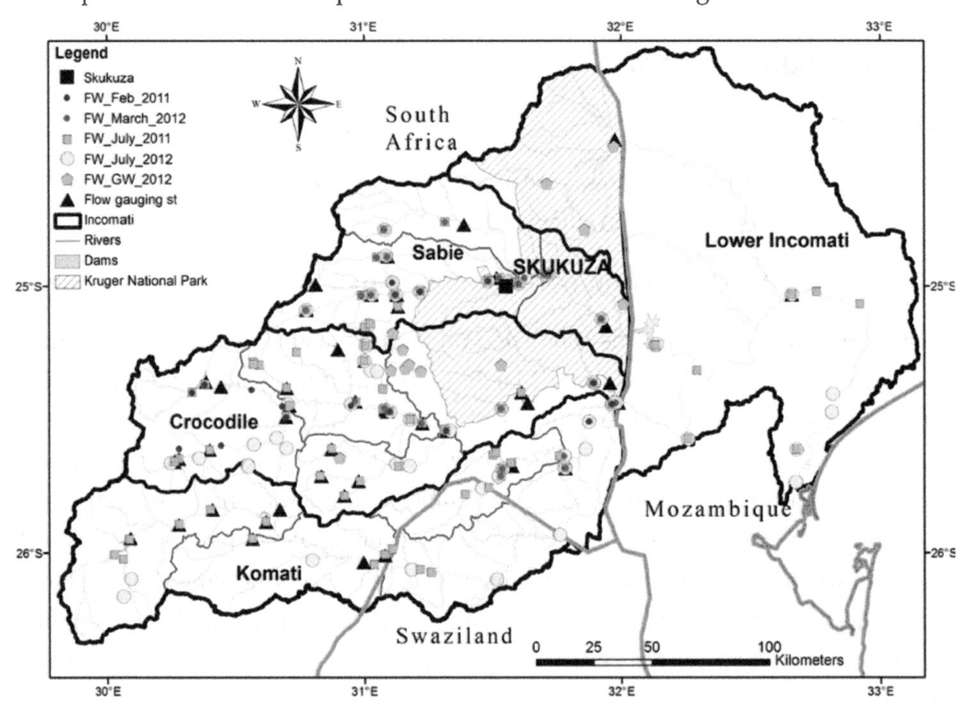

Figure 4-1. Incomati River Basin, main sub-catchments, selected gauging stations and sampling locations for the different fieldwork campaigns

4.2.3 Collection and analysis of samples

Sampling was mainly done from bridges because of ease of access. Water was collected from the river using a bottle with weights and a rope. Sample bottles were washed twice before the final sample was retained and in situ measurements were taken. All the samples were clearly and uniquely labelled and kept in a cooler box. From the cooler box, the samples were kept in a fridge. Analysis for isotopes, cations and anions was done within a week from collection. Oxygen and hydrogen isotopic ratios were measured using the Liquid-Water Isotope Analyser from Los Gatos Research (LGR) at UKZN Laboratory. Oxygen and hydrogen compositions were reported as delta values (δ). The measurement is based on high-resolution laser absorption spectroscopy. Total Alkalinity was measured by potentiometric titration. Cations were measure using Varion 700 ICP Optical Emission Spectrometer. Nitrate was measured using Hach DR2000 photometer.

4.2.4 Analysis of water quality data from DWS-WMS database

A comprehensive analysis of meta-data was conducted to the dataset of water quality for the Inkomati basin (X primary catchment), from the Department of Water and Sanitation-Water Management System (DWS-WMS) (http://www.dwaf.gov.za/iwqs/wms/data/). The metadata includes location description (latitude, longitude, features and type of location), median EC, number of samples analysed, first and last dates of sampling and information about the quaternary catchment and closest flow gauging station. This analysis allowed the selection of the locations for fieldwork sampling, as well as overall patterns and trends of water quality at secondary and tertiary catchment level, to further strengthen findings from fieldwork. The water quality parameters available from the database of surface water locations are: Calcium (Ca), Chloride (Cl), Dissolved Major Salts (DMS), Electrical Conductivity (EC), Fluoride (F), Potassium (K), Magnesium (Mg), Sodium (Na), Ammonium Nitrogen (NH4-N), Nitrate + Nitrite Nitrogen (NO3+NO2-N), pH, Ortho Phosphate as Phosphorus (PO4-P), Silicon (Si), Sulphate (SO4) and Total Alkalinity as Calcium Carbonate (TAL).

4.3 Results and Discussion

4.3.1 Spatial overview of water quality from snapshot sampling

The main observations that can be drawn from the sampling exercises and set the benchmark for the catchment hydro-chemical status are presented below (Table 4-1 and Figure 4-2). In both low and high flow seasons, Komati and Crocodile Rivers had higher EC (average 278±105 and 131±65µS/cm, respectively) than Sabie-Sand (average 63±19 µS/cm). Major ions followed a similar trend, but the Lower Incomati (within Mozambique) registered a higher concentration of alkalinity, Calcium, Magnesium, Sodium and Silica. pH did not register a significant variation; temperature was low during the low flows, which occur in winter and ORP was high for the same period, but with no significant variation.

Table 4-1. Summary of the results indicating mean values of the studied variables and their variation (Standard deviation given in parenthesis) in the Incomati basin.

Parameter		Komati	Crocodile	Sand-Sabie	Lower Incomati	Incomati Basin
EC [μS/cm]	February	277.7 *(105.1)*	131.1 *(65)*	63.3 *(18.9)*	na na	132.4 *(85)*
	July	100.7 *(49.4)*	107.8 *(69.3)*	72.1 *(27.4)*	177.6 *(103.7)*	99.1 *(61.2)*
Temperature [°C]	February	27.4 *(0.2)*	24.7 *(1.3)*	25.9 *(1.2)*	na na	25.3 *(1.5)*
	July	15.6 *(3.1)*	13.7 *(2.5)*	16.7 *(2.7)*	20.2 *(0.6)*	15.6 *(3.1)*
ORP [mV]	February	134.0 *(12.7)*	127.2 *(24.1)*	147.0 *(25.6)*	na na	132.6 *(23.8)*
	July	223.3 *(32)*	218.7 *(50.9)*	104.9 *(70)*	na na	170.7 *(86.4)*
pH [-]	February	7.8 *(0.1)*	8.0 *(0.2)*	7.7 *(0.1)*	na na	7.9 *(0.2)*
	July	6.3 *(0.5)*	6.6 *(1.5)*	7.2 *(0.5)*	7.5 *(0.3)*	6.7 *(1)*
δ^2H [‰]	February	-6.10 *(0.88)*	-9.84 *(1.38)*	-9.14 *(1.63)*	na na	-9.23 *(1.79)*
	July	-6.19 *(3.42)*	-8.39 *(4.53)*	-8.57 *(4.62)*	-2.38 *(2.79)*	-7.49 *(4.45)*
$\delta^{18}O$ [‰]	February	-2.22 *(0.02)*	-2.85 *(0.27)*	-3.37 *(0.42)*	na na	-2.90 *(0.44)*
	July	-2.68 *(0.74)*	-2.92 *(0.88)*	-2.95 *(0.87)*	-1.73 *(0.33)*	-2.79 *(0.86)*
Alkalinity [mg/L]		81.5 *(34.7)*	67.9 *(34.3)*	46.3 *(69.6)*	87.2 *(26.5)*	65.9 *(49.4)*
Calcium [mg/L]		9.2 *(4.3)*	11.1 *(6.4)*	8.6 *(9.3)*	18.3 *(3.7)*	10.3 *(7.2)*
Magnesium [mg/L]		7.8 *(3.7)*	9.1 *(6.5)*	5.5 *(8.1)*	13.4 *(3.3)*	7.9 *(6.6)*
Sodium [mg/L]		10.2 *(7)*	11.3 *(8.4)*	18.3 *(49.7)*	29.7 *(4.7)*	14.3 *(28.6)*
Potassium [mg/L]		0.2 *(0.1)*	0.2 *(0)*	0.2 *(0.1)*	0.3 *(0)*	0.2 *(0.1)*
Silica [mg/L]		7.7 *(1.8)*	8.1 *(2.9)*	7.2 *(2.5)*	10.2 *(2.6)*	7.9 *(2.6)*
Nitrate [mg/L]		0.0 *(0)*	1.4 *(1.3)*	1.1 *(1.2)*	2.6 *(2.3)*	1.0 *(1.3)*

In terms of stable isotopes, Deuterium (^2H) seemed to increase from -14‰ on the headwaters to -4‰ downstream particularly along the Crocodile and Sabie Rivers during high flows. Oxygen 18 (^{18}O) had smaller variations (-2.9±0.9‰), but also tends to be slightly higher downstream in the Crocodile. This pattern of enrichment was observed also in the Gariep River in South Africa (Diamond and Jack, 2018), where ^{18}O varied from -3 to 1‰ from source to sea (over 2000 km). These values of isotopes are similar in range to other studies conducted in South Africa (Wenninger *et al.*, 2008; Abiye *et al.*, 2013; Riddell *et al.*, 2013). In both high and low flow seasons the Komati waters seem to be more enriched in heavy isotopes than the Crocodile River, suggesting a greater evaporative enrichment from irrigation return flows and storage. The Sabie River appears to have the most isotope depleted waters suggesting that evaporative enrichment as a result of land-use activities in this catchment is of low significance. The general grouping for the low flows in the Crocodile River is suggestive of a sustained relatively undistributed water source, which would be explained by the traversing of this river course through dolomite regions in its middle reaches. A local meteoric water line was derived for Skukuza (location indicated in Figure 4-1), and we can see that majority of the samples plot close this line (Figure 4-3).

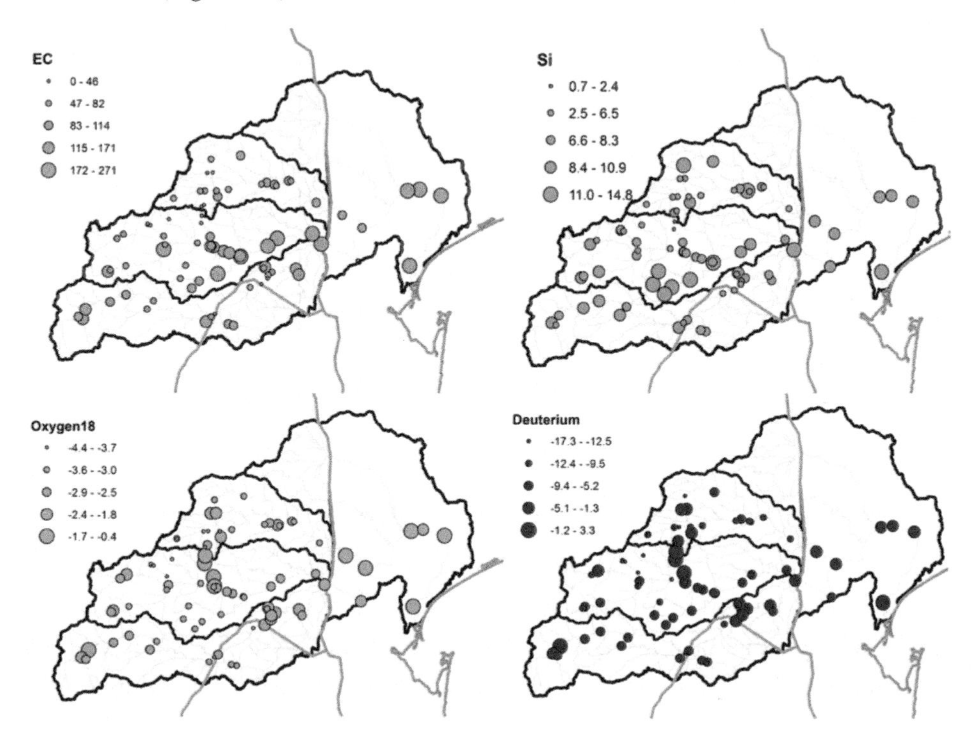

Figure 4-2. Snapshot sampling results for July 2011 for the parameters Electrical conductivity (EC), Silica (Si), Oxygen 18 and Deuterium.

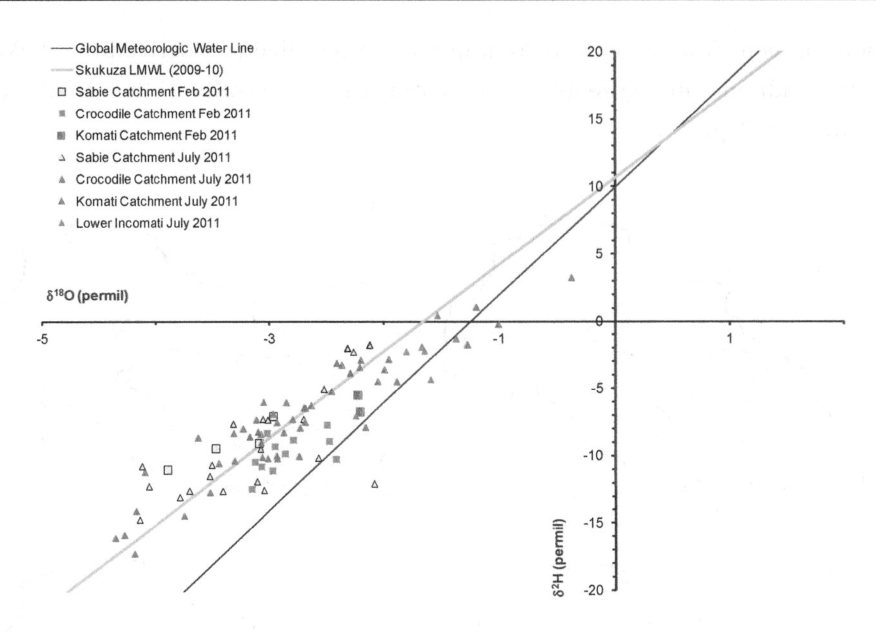

Figure 4-3. Stable Isotope distributions plotted per sub-catchment of the Incomati for 2011

Another relevant finding is the effect of scale on the water quality (Figure 4-4), and the impact of dilution of Sabie River on the Incomati waters. The Sabie's waters are much more pristine than Komati and Crocodile Rivers, so after the confluence with these the overall EC of Incomati reduces, as a result of the mixing and dilution.

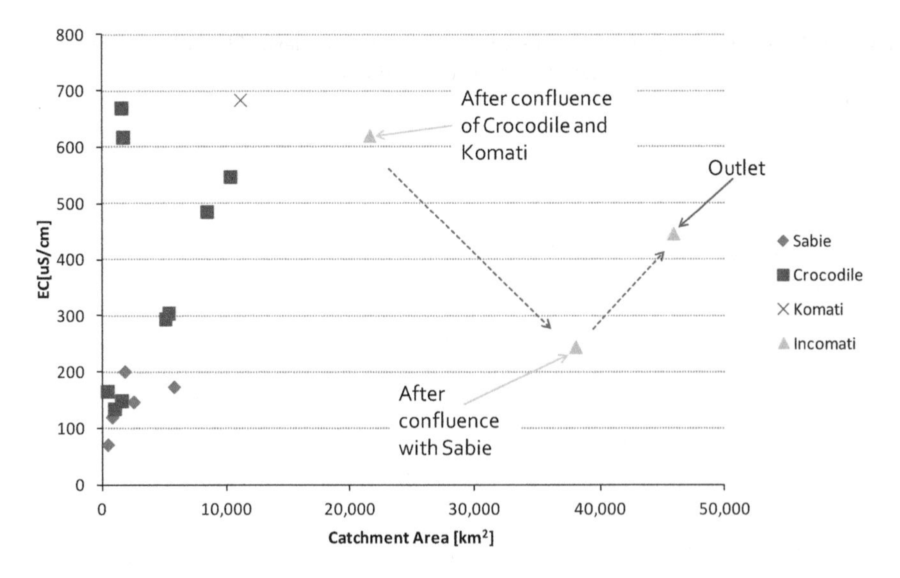

Figure 4-4. Scaling effects of EC as a function of the drainage area, for the main sub-catchments of Incomati

4.3.2 Analysis of water quality dataset from DWS-WMS

The dataset of water quality for the Inkomati (DWS-WMS) includes 339 sites of surface water and 939 sites of groundwater. From the surface water sites 33% (112) are located in the Komati catchment, 47% (160) in the Crocodile and 19% (66) in the Sabie. However, a number of the sites are no longer monitored regularly; the average number of samples per site is 164 ± 276 (average ± standard deviation; the number of samples collected per site range from 1 to 2037, in 2017). Most intense monitoring occurs in river sites, in the Komati and Crocodile catchments. Table 4-2 presents the breakdown of surface water site types, per secondary sub-catchment in the Inkomati catchment.

It is also observed that sampling frequency reduced over time. From 1970s to 1990s the sampling frequency varied between weekly, biweekly to monthly. From 2000s the frequency dropped to mostly monthly. And since 2008 most locations have only few samples collected quarterly, or once every two months.

Table 4-2. Number of surface water sampling sites per type in the Inkomati (DWS-WMS South Africa) in 2017

Secondary Catchment	Canal	Dam	Industry	PWTW	Pipeline	Rivers	Spring/Eye	Unknown Transfer	WWTW	unknown
X1 - Komati	5	12	0	3		81	1	0	6	4
X2 - Crocodile	2	7	1	2	1	126	5	1	12	3
X3 - Sabie	1	4	0	0	2	47	2	2	4	4
X4	0	0	0	0	0	0	0	0	1	0
Total Inkomati	**8**	**23**	**1**	**5**	**3**	**254**	**8**	**3**	**23**	**11**

Out of the 939 groundwater sites, only 8 have more than 10 samples. The average number of samples is 1.7 ± 3.7 (average ± standard deviation), which indicates that most boreholes were only sampled once, and are not regularly monitored. The borehole's distribution is 11% in the Komati, 13% in the Crocodile, 67% in the Sabie and 8% in the X4 catchments.

Table 4-3 shows mean, standard deviation and number of samples of ECmed reported per site, aggregated by secondary catchment and site type. It is noticeable that in the Komati catchment Potable Water Treatment Works (PWTW) has the highest average EC, followed by springs and Waste Water Treatment Work (WWTW) sites. This is the case because two sites categorized as PWTW should be WWTW (X13 196200 and X23193026); One of the sites has very high average EC (169 mS/m), which influences the overall average EC of PWTW. The river water average is 38.2 mS/m, but there is a wider variation on the range of values recorded (Figure 4-5). In the Crocodile catchment, the highest average EC is recorded in a pipeline site (92 mS/m), followed by industry (76 mS/s), transfers (75 mS/s) and springs (73.2mS/s). The pipeline site is X24 192628 "Komatipoort WWTW at Final Discharge at the Last Point Exit from the Plant", so it should be categorized as WWTW. Several of the sites categorized as transfer sites are outlets of WWTW and sewage works. The river water averages 32 mS/m, but a much higher number of samples is collected under this category, with a much wider range (Figure 4-6) so this can significantly affect the mean reported. There are a much higher number of sites affected by anthropogenic activities (industry, agricultural return flows, sewage discharges, mining), and the EC is generally higher than in other catchments (Komati and Sabie), revealing the strong influences of human activities in this catchment. In the Sabie catchment the EC is generally lower than in the other catchments, with average river EC of 14.9 mS/m, reflecting its more pristine conditions. Detailed distribution of mean EC per site type is illustrated in Figure 4-7.

Table 4-3. Average EC (mS/m) per secondary catchment and surface water site type

	ECmed								
Type\Secondary Catchment	mean			std			Total of samples		
	X1	X2	X3	X1	X2	X3	X1	X2	X3
Canal	20.6	29.5	5.0	10.9	37.5	-	314	5	191
Dam	15.8	9.8	13.7	6.7	5.7	15.0	2781	1305	732
Industry	-	76.0	-	-	-	-	-	3	-
PWTW	99.0	58.5	-	83.4	41.7	-	238	76	-
Pipeline	-	92.0	35.0	-	-	1.4	-	69	103
Rivers	38.2	32.0	14.9	44.7	32.6	9.5	9071	27933	10131
Spring/Eye	61.0	73.2	15.0	-	53.9	11.3	50	57	104
Unknown Transfer	-	75.0	39.0	-	-	4.2	-	40	105
WWTW	58.2	68.3	32.8	17.2	20.8	9.0	401	1238	216
unknown	40.8	17.3	34.8	23.3	15.3	34.2	70	113	67

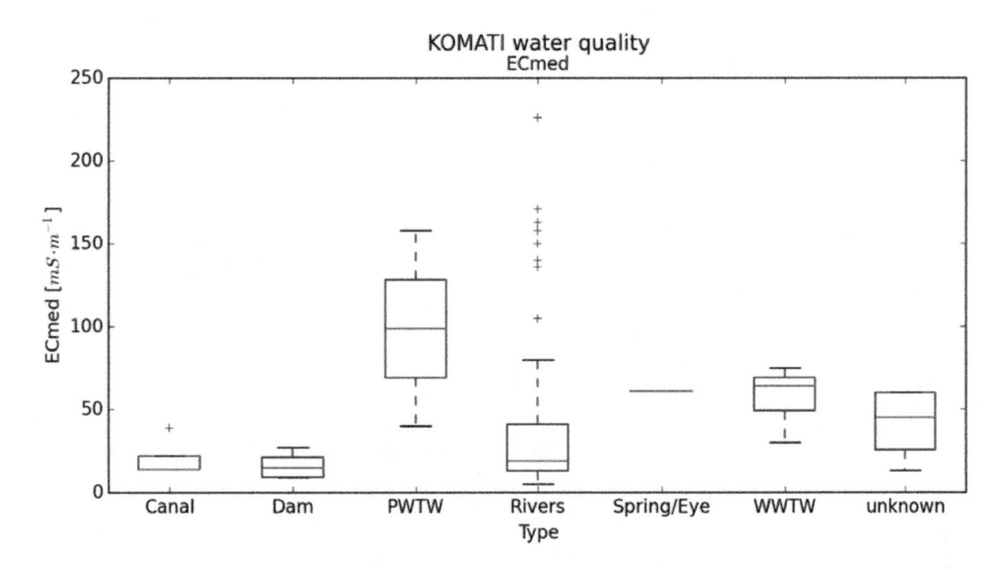

Figure 4-5. Boxplot of ECmed per site type in the Komati catchment.

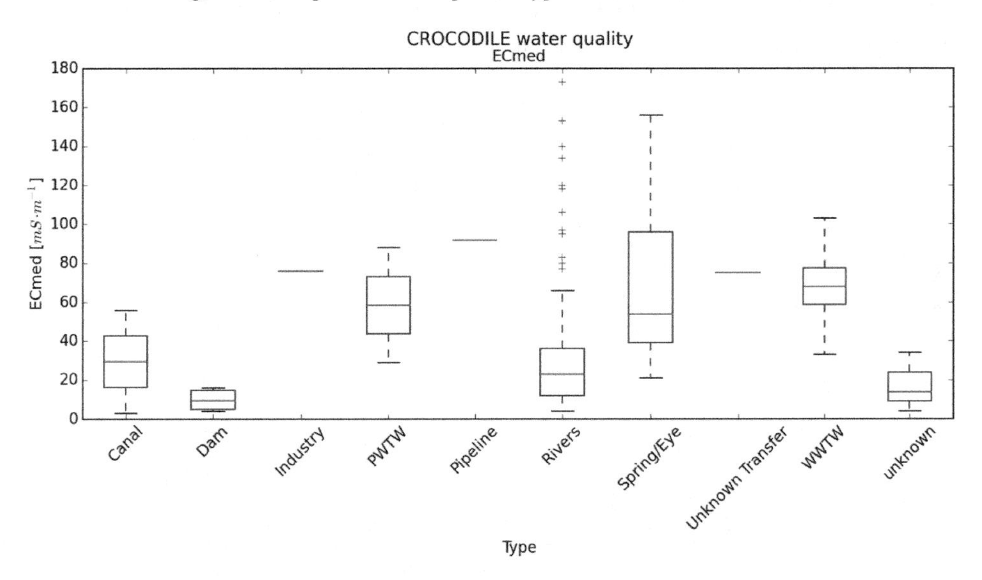

Figure 4-6. Boxplot of ECmed per site type in the Crocodile catchment.

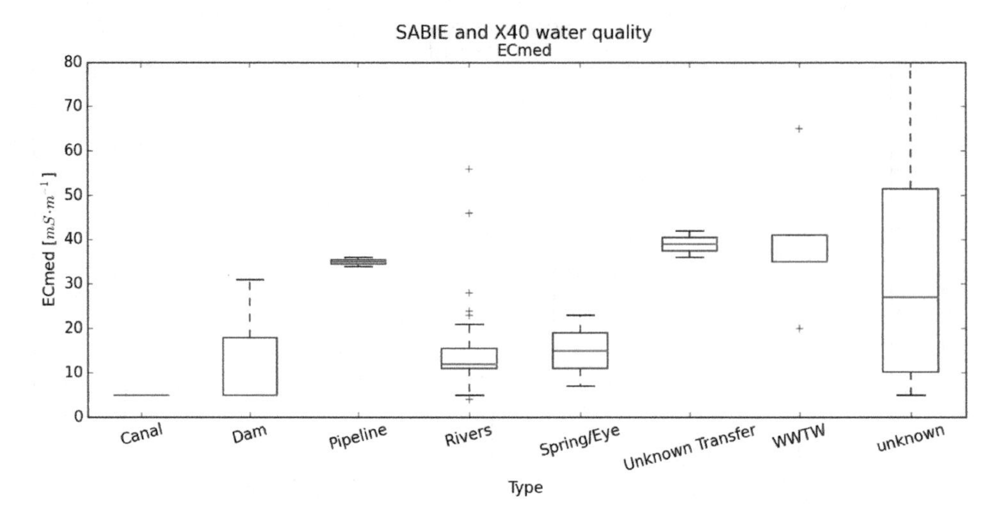

Figure 4-7. Boxplot of ECmed per site type in the Sabie and X40 catchment.

The groundwater ECmed is on average similar across the catchments, but with wider variation in the Sabie, Komati and X4 catchments (Figure 4-8).

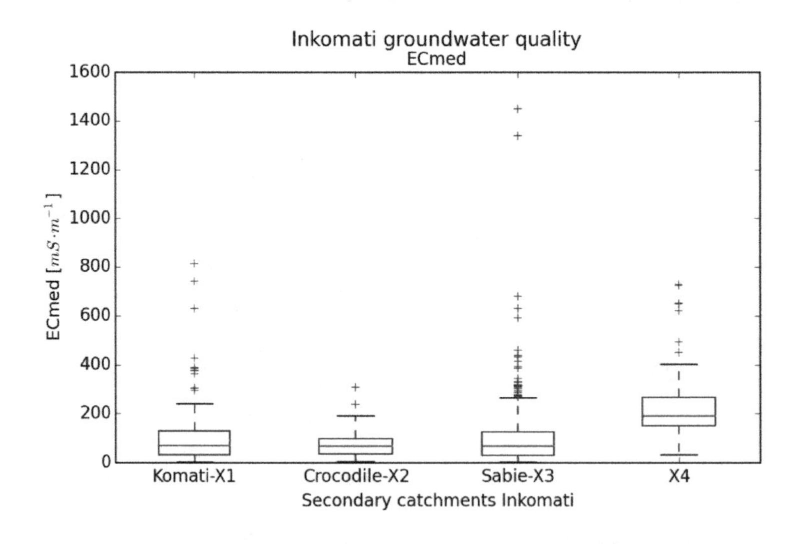

Figure 4-8. Boxplot of ECmed of groundwater per secondary catchment in the Inkomati

4.3.3 Selected river profile stations

From the DWS-WMS database, 16 locations (Figure 4-9 and Table 4-4) were chosen for more detailed analysis of trends over time, and also to map the river profile of main secondary catchments. These locations are also part of the National Chemical Monitoring Programme for Surface Water (NCMP) network (South Africa).

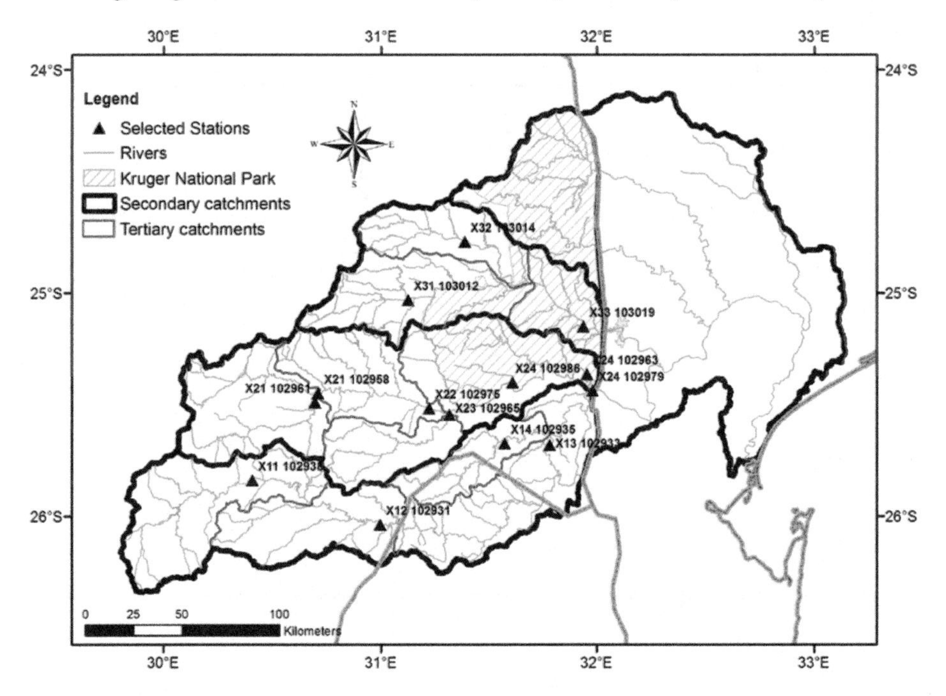

Figure 4-9. Map showing selected water quality stations and gauges for temporal analysis and trends from DWS-WMS database.

The historical data confirm the trend observed in the snapshots, of increase of EC from upstream to downstream the river profile. But in the Crocodile River we can identify the impact of high loads of EC coming from tributaries (Elands and Kaap) into the main river course. A comparison was also made with the water quality standard guidelines of South Africa, to see whether the water quality was complying with the categories of domestic (DWAF, 1996a), irrigation (DWAF, 1996c) and industry (DWAF, 1996b) recommended thresholds (Figure 4-10). It can be observed that the water quality in terms of EC standard for domestic water (70mS/m) is generally adhered to. Only in few locations in the Crocodile and Komati catchments (eg. Tonga, Lomati, Elands, Kaap, Ten Bosch, Komatipoort) the average EC is sometimes higher than the stipulated ideal standard. However, the standards for irrigation (40 mS/m) and industry category 1 (15mS/m) are often exceeded in all locations. Similar pattern was observed for the Chloride standard (results not presented).

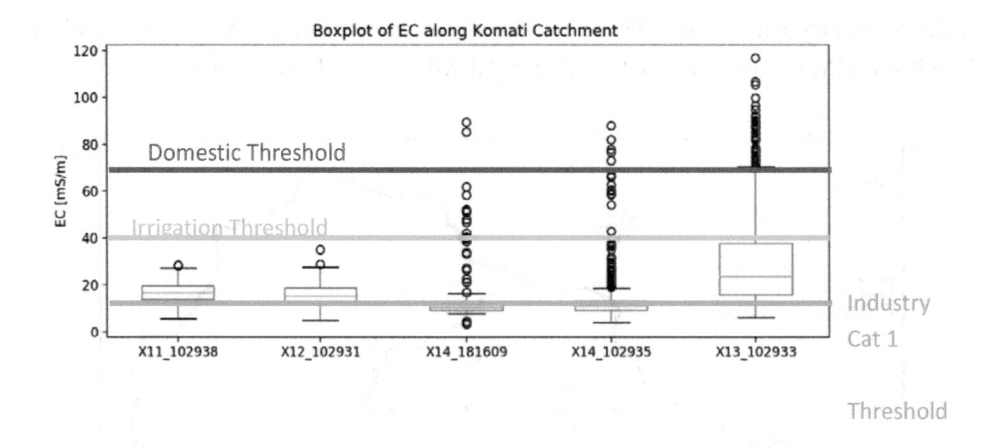

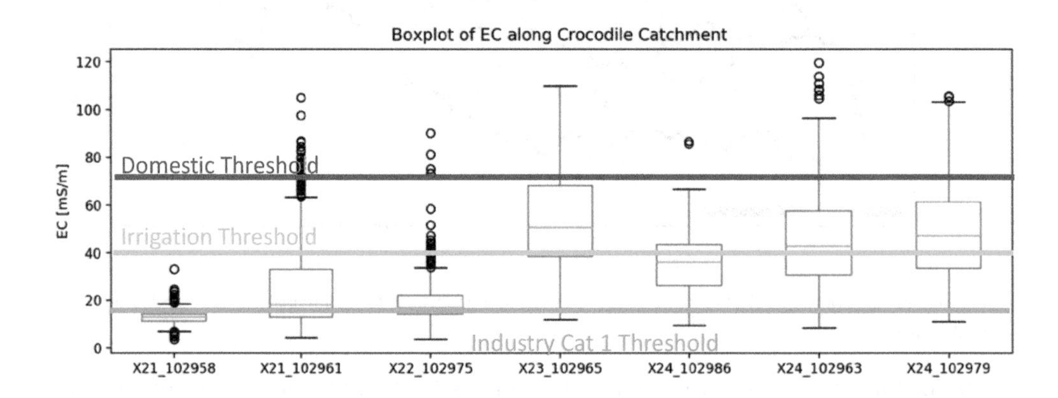

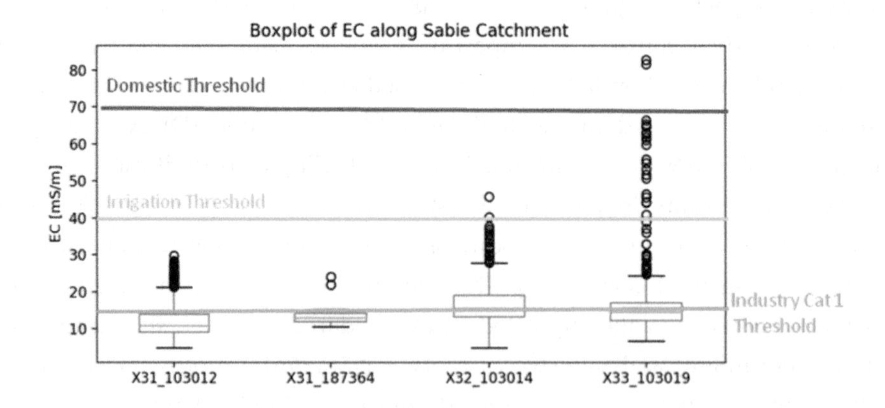

Figure 4-10. Boxplots of ECmed in selected stations along the profiles of Komati, Crocodile and Sabie catchments. For location of the stations refer to Figure 4-9 and Table 4-4.

4.3.4 Temporal variation and trends of water quality

Over the monitoring period, it could be observed that the water quality followed the pattern of dilution during the rainy season, and increased concentrations in the dry season, with highest concentrations in August/September (see for example Figure 4-11, Tonga station in the Komati catchment).

Table 4-4. Selected water quality stations, flow gauges and count of observed trends (significant trends using Mann Kendall test, and $\alpha = 0.05$) in water quality parameters (Ca, Cl, DMS, EC, F, K, Mg, Na, NH4-N, NO3+NO2-N, pH, PO4-P, Si, SO4 and TAL).

					Trends		
	WQ station	Flow gauge	Location	N (years)	Increasing	Decreasing	No trend
Komati	X11_102938	X1H018	Gemsbokhoek	38	12	1	2
	X12_102931	X1H001	Hooggenoeg	36	12	1	2
	X13_102933	X1H003	Tonga	41	12	0	3
	X14_102935	X1H014	Lomati	40	10	0	5
Crocodile	X21_102958	X2H013	Montrose	42	5	1	9
	X21_102961	X2H015	Elands	41	11	0	4
	X22_102975	X2H032	Weltevrede	41	11	1	3
	X23_102965	X2H022	Kaap	39	3	1	11
	X24_102963	X2H016	Ten Bosch	40	5	1	9
	X24_102979	X2H036	Komatipoort	35	8	0	7
	X24_102986	X2H046	Riverside	32	7	1	7
Sabie	X31_103012	X3H006	Perry's Farm	35	12	0	3
	X32_103014	X3H008	Sand	38	10	0	5
	X33_103019	X3H015	Lower Sabie	34	8	1	6

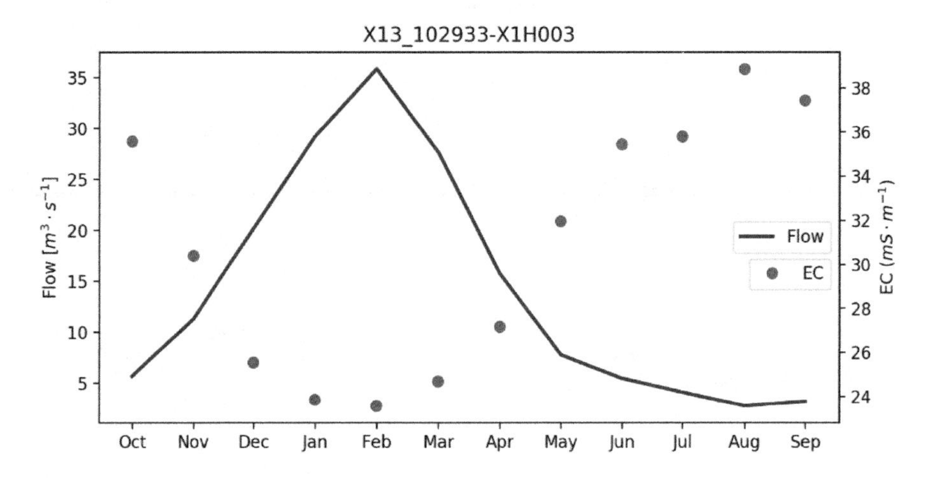

Figure 4-11. Long term mean monthly flow and EC at Tonga station (X13 102933), Komati catchment

Figure 4-12 shows the temporal variation of EC for the selected locations. It is clear the pattern of increase EC from upstream to downstream, and also more marked seasonality at downstream locations. In the Komati catchment the thresholds for irrigation and industry and frequently exceeded at the Tonga and Lomati stations. In the Crocodile catchment the situation is even more critical, with most locations having EC above the industry threshold, and the Elands, Kaap, Ten Bosch, Riverside and Komatipoort have several incidents of EC above the domestic threshold as well.

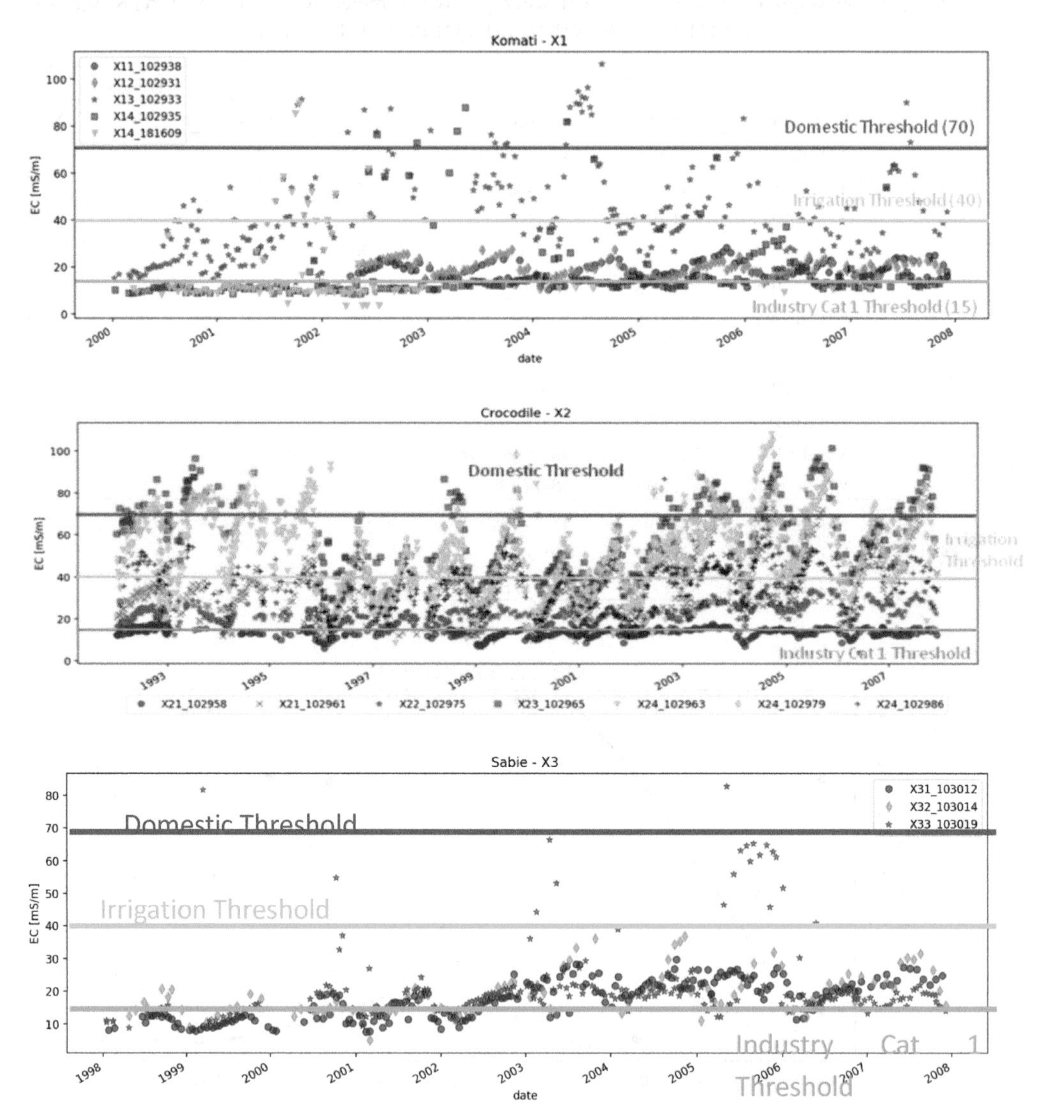

Figure 4-12. Temporal variation of EC in selected stations of the Komati, Crocodile a Sabie catchments.

Most of locations exhibit increasing trends in most water quality parameters over time (Table 4-4). These trends of increased concentration are likely associated with the trends of decreasing flow over the years, particularly in the low flow season, observed in the catchment (Chapter 3). The only decreasing trends were observed for Silica. The parameters that showed no trend are mostly NH_4, NO_3, PO_4, SO_4 – but this varied per location.

A correlation of water quality parameters and flow was conducted for all selected stations. Figure 4-13 shows the results for the Tonga station as an example, which is representative of the results found in other locations. We could identify strong correlations between Ca, Cl, DMS, EC, Mg, Na, TAL and weak correlations between NH_4, NO_3, PO_4, Si. Most water quality parameters had negative correlations with flow, which is explain by the dilution that occurs during high flows, and higher concentrations observed during low flows.

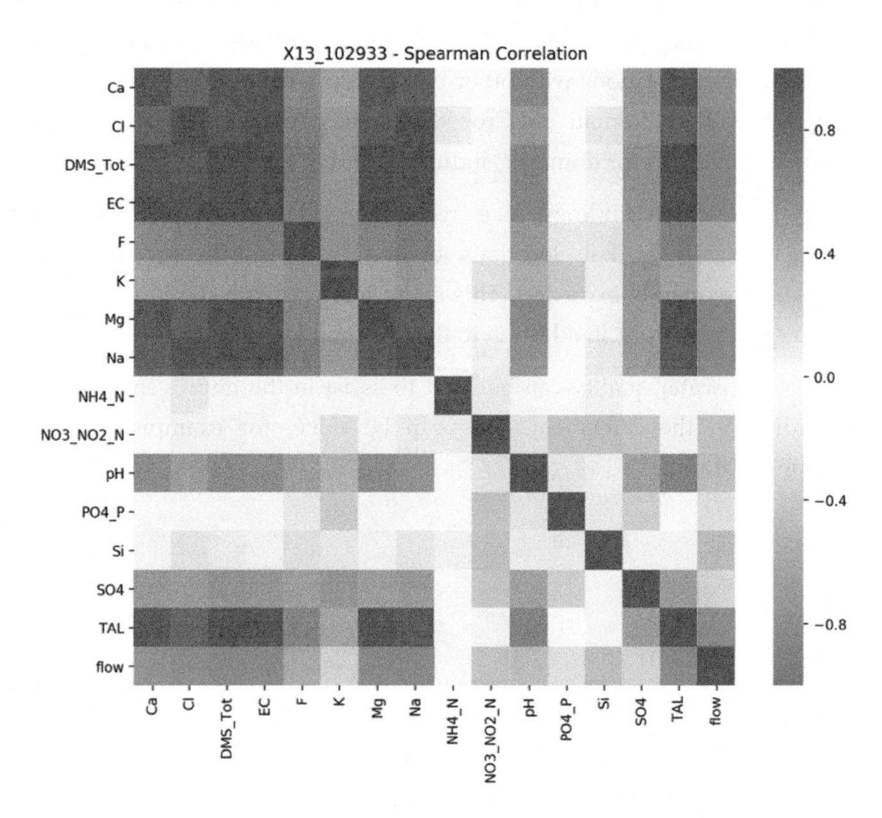

Figure 4-13. Correlation matrix of water quality parameters and flow for the Tonga station, Komati; the bar on the right gives the colour scale of the correlation coefficient.

4.4 Conclusions

The analysis of primary water quality data collected during snapshots and of secondary water quality data revealed several patterns and trends. There is a general increasing trend of major cations, anions, EC from upstream headwaters to downstream in the Incomati Basin. The stable isotopes also indicate more depleted water in headwaters (resembling more rain water) and more enriched water downstream, particularly on Crocodile and Komati systems. In the Crocodile system in particular, Elands and Kaap contribute with quite high EC to Crocodile main stem, with major concern during low flows (May to September).

The presence of irrigation return flows can be detected by elevated EC, high Sodium (Na^{+}) and enrichment of isotopes. The impact of reservoirs can be detected through enrichment of isotopes, due to the process of evaporation.

The analysis of DWS-WM dataset revealed that there was a great expansion in spatial coverage of sampling locations over the years, but there is also an alarming decrease in sampling frequency since 2008. EC is the most regularly monitored parameter, which has strong correlations with other parameters such as major cations. In many locations, especially on Komati and Crocodile, some parameters (EC, Cl, Ca, Na, SO_4) are above SA guidelines for domestic, industrial and other uses.

Based on the research findings, it is recommended that sampling frequency is increased, especially at key locations of the catchment, where water quality thresholds are frequently exceeded. This could be done with the installation of real time water quality sensors, in addition to the flow gauges.

The database of water quality can be used to assist in the improvement of process understanding in the catchment. This can be done, for example, by chemical hydrograph separation.

5

HYDROGRAPH SEPARATION USING TRACERS AND DIGITAL FILTERS TO QUANTIFY RUNOFF COMPONENTS

This chapter explores the use of tracers and water quality to quantify runoff components in a selected sub-catchment of the Incomati, the Kaap catchment. First, spatial and temporal variability of water quality in the catchment is described, and suitable data for hydrograph separation is identified. Then, chemical hydrograph separation is applied using electrical conductivity as a tracer. Furthermore, a recursive digital filter is applied to quantify baseflow using only daily streamflow data. The digital filter parameters are then calibrated using chemical hydrograph separation as a reference. Baseflow and quickflow components are then quantified at monthly and annual scale, using the calibrated digital filter approach. This information is then used to improve understanding of how baseflow is generated and how it contributes to streamflow throughout the year. Implications of this finding for water management are also discussed.

This chapter is based on: Saraiva Okello AML, Uhlenbrook S, Jewitt GPW, Masih I, Riddell ES, Van der Zaag P. 2018b. Hydrograph separation using tracers and digital filters to quantify runoff components in a semi-arid mesoscale catchment. *Hydrological Processes*, 32: 1334-1350. DOI: doi:10.1002/hyp.11491.

5.1 Introduction

Hydrological processes, particularly runoff generation, influence water quality and water quantity as well as their spatial and temporal dynamics. Detailed understanding of these processes is thus important for the prediction of water resources yield, floods, erosion, as well as solute and contaminant transport (Hrachowitz *et al.*, 2011; Blöschl *et al.*, 2013b). The recently concluded IAHS decade on Prediction in Ungauged Basins (PUB) (Hrachowitz *et al.*, 2013) and the current decade, "Panta Rhei-everything flows" (Montanari *et al.*, 2013) stress the importance of making better use of all available data and combining different methods in order to get a more holistic understanding of catchment functioning and hydrological processes.

Semi-arid systems, which constitute a large part of the African continent, are characterized by high spatio-temporal variability of rainfall, high evaporation rates, deep groundwater resources and poorly developed soils (Hughes, 2007; Wheater *et al.*, 2008; Hrachowitz *et al.*, 2011; Blöschl *et al.*, 2013b). Additionally, semi-arid systems have marked seasonality and are more prone to flood and drought extremes, with dry spells that can last for years (Love *et al.*, 2010a). High-intensity storms may generate most of the season's runoff (Love *et al.*, 2010a; Van Wyk *et al.*, 2012). These characteristics, combined with the multiple other challenges such as growing population, increasing water use and irrigation, urbanization, and pollution, make it extremely important to better understand runoff generation in semi-arid areas (Hughes, 2007).

There is no standard procedure to best understand runoff generation processes (Beven, 2012), but several methods are often applied to quantify runoff components, including graphic hydrograph separation, rainfall-runoff models, baseflow filters, and chemical/isotope (tracer) hydrograph separation (Uhlenbrook *et al.*, 2002; Gonzales *et al.*, 2009; Zhang *et al.*, 2013; Cartwright *et al.*, 2014). Hydrograph separation is commonly applied to quantify different components contributing to river flow. River discharge following a rainfall event may be divided into quick flow and baseflow (Hall, 1968; Tallaksen, 1995; Stewart, 2015). Quick flow is the water that contributes to river flow immediately after the rainfall event, but it can include water from different sources (Hrachowitz *et al.*, 2011; Cartwright *et al.*, 2014), such as rainfall, overland flow, older water displaced from unsaturated or saturated zone (Sklash and Farvolden, 1979; Cartwright *et al.*, 2014). Baseflow is water with longer response time to precipitation in the catchment which sustains the river between rainfall events and seasons. It can include contributions from regional groundwater, interflow, water from river banks and floodplains (Sklash and Farvolden, 1979; Tallaksen, 1995; Cartwright *et al.*, 2014). Therefore, these components can be made up of water from different sources, the contributions of which vary in space and time, and further complicate the understanding of underlying processes (Cartwright *et al.*, 2014).

Hydrological process studies are costly and labour intensive (Hughes *et al.*, 2003; Wenninger *et al.*, 2008; Hughes *et al.*, 2015). In addition, there is often limited capital and human resources to perform them, particularly in the poorly gauged catchments of Africa (Hughes *et al.*, 2003; Hughes *et al.*, 2015). The use of tracers in such situations has been shown to be a cost effective and a pragmatic approach. Indeed, several recent studies show the potential of using tracers to improve the understanding of runoff generation processes and flow paths in small catchments (Tetzlaff and Soulsby, 2008; Wenninger *et al.*, 2008; Birkel *et al.*, 2010; Capell *et al.*, 2011; Capell *et al.*, 2012; Birkel *et al.*, 2014b). However, such studies at a larger scale (e.g. $>10^3$ km^2) are still rare (Frisbee *et al.*, 2013; Miller *et al.*, 2014; Miller *et al.*, 2015; Tetzlaff *et al.*, 2015), as many factors influence water quality at the catchment scale, such as geology, soils, elevation, topography, climate, seasonality and land use management.

For hydrograph separation to be useful, tracers should be conservative and not react during their medium of transport through the catchment. This is mostly the case with environmental tracers, such as deuterium and oxygen-18 (Klaus and McDonnell, 2013). However, historical data on the concentration of these conservative tracers in the catchment are generally not available; therefore, hydrochemical tracers are used for hydrograph separation. Several studies (Uhlenbrook *et al.*, 2002; Soulsby *et al.*, 2004; Tetzlaff and Soulsby, 2008; Capell *et al.*, 2011; Miller *et al.*, 2014; Miller *et al.*, 2015) have used Electrical Conductivity (EC), Alkalinity, Chloride, and Silica as tracers to perform hydrograph separation, despite some limitations. Authors argue that different tracers contain different information, for instance Silica can be a good tracer of geographical sources of runoff (Uhlenbrook *et al.*, 2002), whereas Alkalinity could be used as a conservative tracer to differentiate between acidic soil-water and more alkaline groundwater (Tetzlaff *et al.*, 2007; Tetzlaff and Soulsby, 2008). Uhlenbrook *et al.* (2002) used dissolved silica concentration as an indicator of residence time of water, and inferred the source and pathway for runoff components. In that study, overland flow from saturated and impervious areas of the Brugga catchment in Germany were assumed to have no dissolved silica (resembling rainfall) whereas in subsurface flow components the water had time to react with mineral soil and become enriched with dissolved silica depending on the geology and residence time of the water.

A Recursive Digital Filter is a method adapted from signal processing theory (Nathan and McMahon, 1990; Eckhardt, 2005). In application of this method, daily stream flow time series are considered a mixture of quick flow (high-frequency signal) and baseflow (low-frequency signal). By filtering out the high-frequency signals from the streamflow, the low-frequency signals (baseflow) can be revealed. Recursive digital filters have been applied for graphical hydrograph separation (Eckhardt, 2005; Eckhardt, 2008; Cartwright *et al.*, 2014; Miller *et al.*, 2015), using only streamflow records and filter parameters, which can be calibrated with tracer data (Zhang *et al.*, 2013; Rimmer and Hartmann, 2014; Longobardi *et al.*, 2016) or other catchment characteristics (Eckhardt, 2005; Cartwright *et al.*, 2014; Mei and Anagnostou, 2015). Some recent studies have shown the potential of combining hydrochemical tracer, particularly EC, and digital filter methods in order to estimate baseflow for long records of streamflow (Zhang *et al.*, 2013; Li *et al.*, 2014; Miller *et al.*, 2014; Rimmer and Hartmann, 2014; Miller *et al.*, 2015; Longobardi *et al.*, 2016). Hydrograph separation using calibrated digital filters is therefore inexpensive, and allows for quick and objective separation of streamflow components, which are important in the management of water resources. However, this approach has not been tested extensively in arid and semi-arid systems, particularly in Southern Africa.

Building from this body of literature, the main objective of this paper is to use water quality data, especially EC, and a digital filter hydrograph separation method to quantify runoff components in a semi-arid catchment in Southern Africa. EC was chosen as main tracer because it is easy and cheap to measure and widely available, therefore its applicability is important to understand. Depending on geology other tracers like dissolved silica (e.g. Uhlenbrook and Hoeg, 2003) can be useful. However, due to availability this study focused on the examining the potential of EC as tracer.

Thus, the specific objectives are to:

- assess if long term discrete EC data can be used to perform hydrograph separation at the monthly scale in a meso-scale semi-arid catchment;

- perform hydrograph separation using digital filters at daily time scale and assess the methodology to calibrate filter parameters using EC data; and
- quantify the relative contribution of quick flow and baseflow to total runoff at both monthly and annual time scales.

5.2 Methodology

5.2.1 Study area - The Kaap catchment

The Kaap catchment is located in the northeast of South Africa in the Mpumalanga Province and drains an area of approximately 1640 km^2. Flowing from west to east, the Kaap River joins the Crocodile River which then flows to the trans-boundary Incomati River. The Kaap catchment contains three main tributaries: the Queens, the Suidkaap, and the Noordkaap (see Figure 5-1 and Table 5-1). The study area is located in the low elevation sub-tropical region of South Africa, Swaziland and Mozambique known as the Lowveld with elevations ranging from 300 m to 1800 m above sea level (see Figure 5-1A). The climate is semi-arid with cool dry winters and hot wet summers, characterized by a distinct wet and dry season, with significant intra-seasonal variability associated with the summer rainfall season which is dominated by thermal and orographic thunderstorms. The wet season typically runs from October to March, as illustrated in Figure 5-2. Precipitation ranges from 583 mm y^{-1} to 1243 mm y^{-1} in the highest parts of the catchment (Middleton and Bailey, 2009). The mean potential evaporation is estimated to 1435 mm y^{-1}(Middleton and Bailey, 2009).

Streamflow is highly seasonal with the highest average flow occurring in February with an average of 9.2 $m^3 s^{-1}$ at the outlet and a mean annual runoff coefficient of 0.14. The lowest flow during the year is observed at the end of the dry season in September falling to an average of 0.8 $m^3 s^{-1}$. Minimum and maximum mean daily flows recorded between 1961 and 2012 at the Kaap outlet range from 0 (below detection limit) to 483 $m^3 s^{-1}$. The catchment is fairly well monitored with five streamflow gauges available within the catchment (see Figure 1A). According to Bailey and Pitman (2015), the long term natural mean annual runoff of the Kaap River catchment is 189 $\cdot 10^6$ $m^3 y^{-1}$ (equivalent to 116 mm y^{-1}). Table 5-1 describes the physiographic characteristics of the Kaap and tributaries in more detail. Woody savannah (Bushveld) and grasslands are the dominant land cover in the Kaap Valley covering up to 68% of the catchment as observed in Figure 5-1B and Table 5-1. In the upper areas, approximately a quarter of the total catchment consists of exotic pine and eucalyptus plantations used for paper and timber production. Sugarcane, vegetables and citrus orchards are found in the lower part and are irrigated. No other major structures, such as reservoirs, are present in the catchment. The total human population of the Umjindi municipality (which cover most of the Kaap catchment) is over 71200 persons, distributed between the town of Barberton (with a population of over 12000), and several townships, farms and informal settlements (Stats, 2016). There are some gold mines still active in the catchment.

The Kaap valley presents some of the oldest rock formations on Earth, including the Onverwacht group, some 3.5 billion years old (Sharpe *et al.*, 1986; de Wit *et al.*, 2011). Biotite granite is the predominant formation in the valley as observed in Figure 5-1C. The headwaters have weathered granite which has felsic properties indicating high concentrations of silica (Hessler and Lowe, 2006). In contrast, surrounding the granite, lava formations are present in the form of basalt and peridotitic komatiite which are low in silicates and high in magnesium. Sandstones and shales are found in close proximity to the Kaap River and at the south section of the catchment (Sharpe *et al.*, 1986; Hessler and Lowe, 2006). In addition to the gneiss formation observed at the outlet, other formations present include ultramafic (high in iron and low silicates) rocks, quartzite and dolomite (see Figure 5-1C and Table 5-1). Borehole logs near the upper Suidkaap and Noordkaap tributaries displayed a top layer of weathered granite (approximately 25 to 37 m in depth) followed by a thinner, less fractured granite layer and hard rock granite. Borehole logs analyzed in proximity to the catchment outlet presented more diverse formations including layers of clay, sand, greywacke and weathered shale.

The predominant soils are rhodic ferrasols, chromic cambisols and haplic acrisols in the headwater catchments. In the Kaap valley, lithic leptosols and rhodic nitisols dominate (Hengl *et al.*, 2014; Hengl *et al.*, 2017). In terms of soil texture, 53% of the Kaap catchment is covered in sandy clay loams, 39% in clay loams and the remainder of the catchment has clays, sandy clays and sandy loams (Hengl *et al.*, 2014; Hengl *et al.*, 2017).

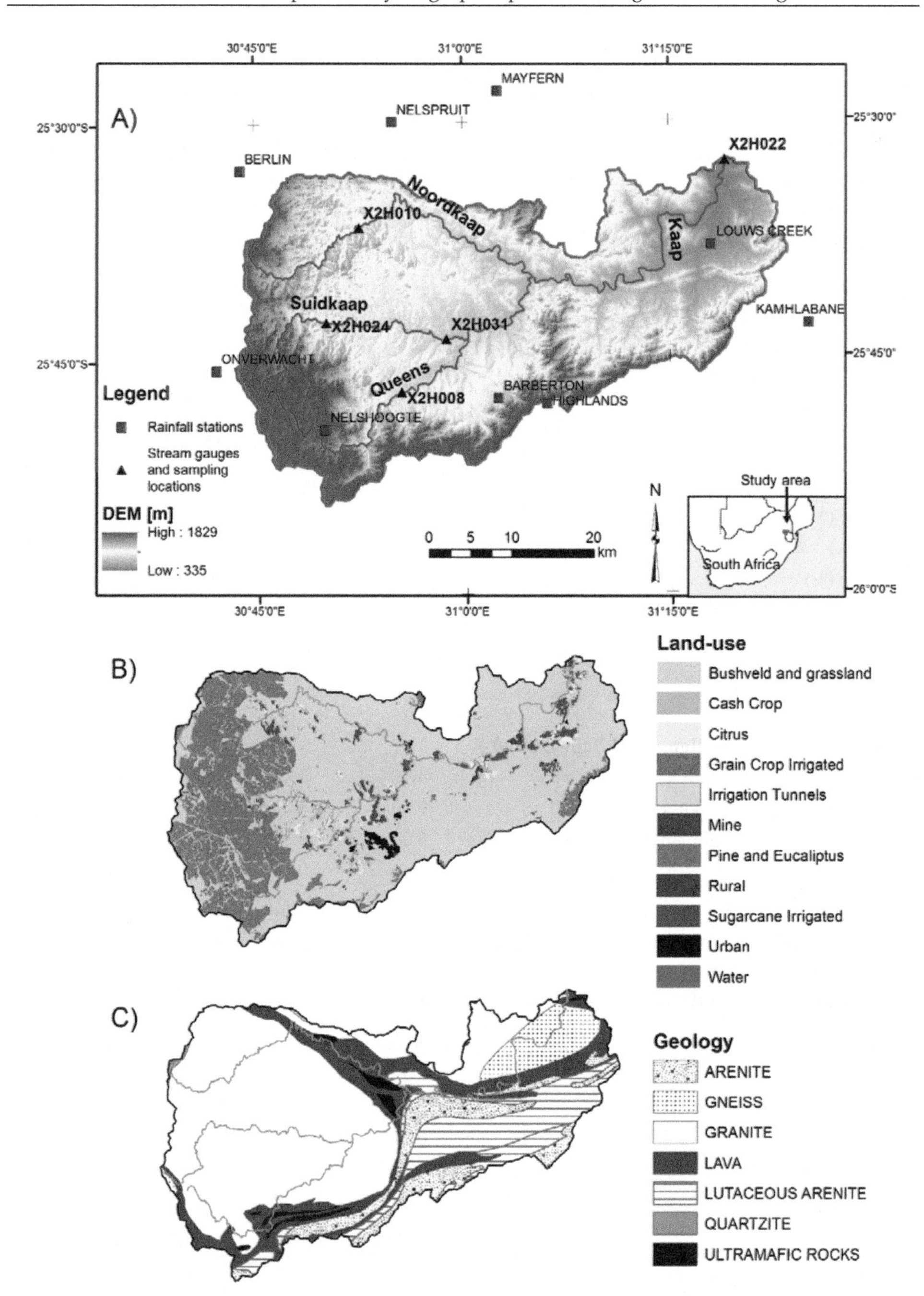

Figure 5-1. A) Location of Kaap catchment in South Africa (inset) and DEM of Kaap catchment with sampling locations, stream gauges and rainfall stations; B) Land-use and land-cover map of Kaap catchment; and C) Geological map (Middleton and Bailey, 2009)

Table 5-1. Catchment physiographic and hydro-climatic characteristics

Tributary Name	Kaap Outlet	Queens	Suidkaap	Noordkaap
Hydrometric Station ID	X2H022	X2H008	X2H031	X2H010
Sub-basin area (km^2)	1640	180	262	126
Reach Length up to station (km)	108	30	41	25
Topography				
Mean elevation (m.a.s.l)	899	1261	985	1090
Min elevation (m.a.s.l)	332	733	651	847
Max elevation (m.a.s.l)	1862	1688	1862	1755
Mean slope (degrees)	11	11	7	9
Max slope (degrees)	60	51	45	57
Geology				
Granite (%)	53	60	98	97
Lava (%)	16	28	1	0
Arenite (%)	9	2	0	0
Ultramafic (%)	2	4	0	0
Quartzite (%)	0	0	0	3
Gneiss (%)	6	0	1	0
Lutaceous Arenite (%)	14	6	0	0
Land use (2004) (Middleton and Bailey, 2009)				
Bushveld and grassland (%)	67.6	36.3	37.1	30.0
Planted forest (pine and eucalyptus) (%)	25.0	63.5	55.6	68.4
Irrigated cultures (sugarcane, citrus, others) (%)	6.0	0.0	7.0	1.6
Urban and mines (%)	1.4	0.1	0.4	0.1
Hydro-meteorology				
Mean Annual Precipitation from 1950-2010 [mm y^{-1}]	900	1016	905	1101
Potential ET (S-Pan) [mm y^{-1}]	1435	1369	1451	1425
Runoff (MAR) natural (WR2012, 1970-2010) [mm y^{-1}]	116	146	210	216
Runoff (MAR) observed (1970-2010) [mm y^{-1}]	66	99	120	149
Qmin (m^3 s^{-1})	0	0	0	0
Q95 (m^3 s^{-1})	0.01	0.02	0.12	0.12
Qmean(m^3 s^{-1})	1.35	0.25	0.61	0.40
Q5 (m^3 s^{-1})	13.18	2.05	2.89	1.58
Qmax (m^3 s^{-1})	482.63	95.88	123.35	27.93

5.2.2 Data used

Hydrological data in the catchment including precipitation, evaporation, streamflow and groundwater records were collected from the South African Department of Water & Sanitation (DWS, former DWA), the South African Weather Service (SAWS)

and the South African Sugarcane Research Institute. Geological, topographical and land use data were obtained from Middleton and Bailey (2009), and the Catchment Management Strategy studies (ICMA, 2010). To analyse the flow behaviour at the outlet and tributaries, average daily discharges at X2H022 (Outlet), X2H008 (Queens), X2H031 and X2H024 (Suidkaap) and X2H010 (Noordkaap) stream gauges were obtained from the DWS. Their locations are shown in Figure 5-1A. Time series of water quality (Electrical conductivity, pH, Calcium, Magnesium, Potassium, Sodium, Chloride, Sulphates, Total Alkalinity, Silica, Fluoride, Nitrates, Ammonia, Phosphate and Total Dissolved Salts) were obtained from the DWS-Water Management System (http://www.dwaf.gov.za/iwqs/wms/data/). The time series of water quality is intermittent, with weekly or fortnightly samples in some years and monthly samples in others. Rainfall water quality was not part of the routine sampling program undertaken by DWS. Table 5-2 details the location, time series length and source of data used.

Table 5-2. Data used for rainfall, flow and water quality

	Code	Name	Latitude	Longitude	Altitude	Start year	End year	Time resolution/ Number of samples	Institution/source
Rainfall	0519733 9	Kamhlabane	-25.717	31.417	1205	1972	2014	daily	SAWS
	0519518 7	Louws Creek-Pol	-25.633	31.300	477	1972	2014	daily	SAWS
	0519168 0	Highlands	-25.800	31.100	1240	1995	2014	daily	SAWS
	0519077 8	Barberton - TNK	-25.794	31.041	852	1972	2009	daily	SAWS
	0518589 3	Nelshoogte Bos	-25.825	30.832	1400	1972	2014	daily	SAWS
	0555750 9	Nelspruit	-25.500	30.916	883	1993	2014	daily	SAWS
	0518393 3	Berlin Bos	-25.550	30.733	1341	1972	2014	daily	SAWS
	0556088 4	Mayfern	-25.468	31.043	610	1973	2014	daily	SAWS
	0518346 2	Onverwacht	-25.761	30.701	1403	1972	2014	daily	SAWS
Flow	X2H008	Queens River at Sassenheim	-25.786	30.924		1948	2014	daily	DWS - HS
	X2H010	North Kaap River at Bellevue	-25.611	30.875		1948	2014	daily	DWS - HS
	X2H022	Kaap River at Dolton	-25.543	31.317		1960	2014	daily	DWS - HS
	X2H031	South Kaap River at Bornmans Drift	-25.730	30.978		1966	2014	daily	DWS - HS
Water Quality [1]	X2H008Q01	Queens River at Sassenheim	-25.786	30.924		1978	2012	465	DWS - WMS
	X2H010Q01	North Kaap River at Bellevue	-25.609	30.875		1978	2012	454	DWS - WMS
	X2H022Q01	Kaap River at Dolton	-25.542	31.317		1978	2012	940	DWS - WMS
	X2H031Q01	South Kaap River at Bornmans Drift	-25.729	30.979		1978	2012	479	DWS - WMS

[1] - Water quality parameters include: Electrical conductivity, pH, Calcium, Magnesium, Potassium, Sodium, Chloride, Sulphates, Total Alkalinity, Silica, Fluoride, Nitrates, Ammonia, Phosphate and Total Dissolved Salts

SAWS: Southern Africa Weather Service

DWS – HS : Department of Water and Sanitation (former Department of Water Affairs) – Hydrological Services, https://www.dwa.gov.za/hydrology/

DWS – WMS : Department of Water and Sanitation (former Department of Water Affairs) – Water Management System, http://www.dwaf.gov.za/iwqs/wms/data/

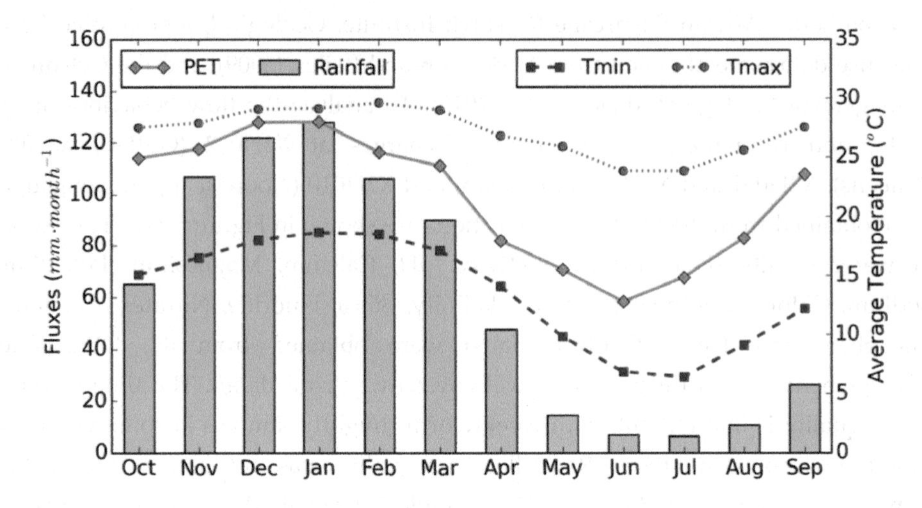

Figure 5-2. Long term monthly average rainfall, potential evaporation, minimum and maximum temperature for the Kaap catchment

5.2.3 Chemical hydrograph separation

Electrical conductivity (EC) data were combined with discharge data to perform a two component hydrograph separation based on steady state mass balance equations of water and tracer fluxes, as described by Equations (5.1) and (5.2).

$$Q_t = Q_s + Q_b \qquad [5.1]$$

$$c_t\, Q_t = c_s\, Q_s + c_b Q_b \qquad [5.2]$$

Where Q_t is mean daily discharge at sampling point [$m^3\ s^{-1}$], Q_s is the runoff contribution from the surface runoff [$m^3\ s^{-1}$] and Q_b is the runoff contribution from the subsurface runoff [$m^3\ s^{-1}$]; c_t is the tracer concentration at the sampling point [$\mu S\ cm^{-1}$];, c_s is the tracer concentration of the surface runoff [$\mu S\ cm^{-1}$] and c_b is the tracer concentration of the sub-surface runoff [$\mu S\ cm^{-1}$].

The concentration for sub-surface runoff was assumed to be the concentration of streamflow when the flow is lowest (assuming these are baseflow conditions) and the concentration of the surface runoff was assumed to be similar to concentrations observed during rainfall events (van Wyk *et al.*, 2011; Camacho Suarez *et al.*, 2015). The assumptions of this method are further discussed by Buttle (1994), Uhlenbrook *et al.* (2002) and Uhlenbrook and Hoeg (2003).

The two component hydrograph separation was performed using EC. The main assumption for the definition of the groundwater end member is that during the dry season, the baseflow is essentially made up of groundwater alone. Thus, the highest percentiles of EC were considered to represent the groundwater signature.

The hydrograph separation was conducted using daily flow data and intermittent EC data, therefore results are compared at monthly and annual scales.

5.2.4 Digital filters hydrograph separation

Hydrograph separation using recursive digital filter approaches focus on distinguishing between rapidly occurring discharge components such as surface runoff, and slowly changing discharge originating from interflow and groundwater (Rimmer and Hartmann, 2014).

The hydrograph separation algorithm given in Equation 5.3 was implemented in Python code, to facilitate repetition, calibration and plotting. The recursive digital filter (Eckhardt, 2005) is based on the assumption that the outflow from an aquifer is linearly proportional to its storage:

$$b_k = \frac{(1-BFI_{max})\alpha b_{k-1}+(1-\alpha)BFI_{max}y_k}{1-\alpha BFI_{max}} \qquad [5.3]$$

Subject to $b_k \leq y_k$, where:

b_k is the baseflow flux on day k, y_k is total discharge on day k, α is the recession constant that is estimated from the recession limbs of the hydrograph, BFI_{max} is the maximum value of the baseflow index (the long term ratio of baseflow to river discharge) that can be modelled by the algorithm.

According to Eckhardt (2005) and Eckhardt (2008) the BFI_{max} is a very sensitive parameter, but is also non-measurable. Therefore, Eckhardt (2008) suggested that tracers can provide data for better calibration of BFI_{max}. The 'BFI_{max}' parameter could be better defined using the independent results of tracer hydrograph separation (Zhang *et al.*, 2013; Cartwright *et al.*, 2014; Rimmer and Hartmann, 2014), which was the approach adopted in this research.

The parameter α was computed using three different methods, described in Rimmer and Hartmann (2014): First, a master recession curve (Eckhardt, 2005) was constructed, where the daily streamflow data of several dry seasons was overlapped based on the day of the year (DOY) for the period of May to September (DOY 121-250). The master recession curve was created by averaging the flow during all seasons, and the exponent that best fitted the recession equation was used to compute α. Second, the correlation method (Tallaksen, 1995) was used, where α corresponds to the slope of the regression line through the origin in a scatter plot of streamflow Q_{j+1} against Q_j. Third, the mean value of Q_{j+1}/ Q_j during the same dry period was calculated.

5.2.5 Calibration of digital filter parameters with tracer data

The calibration procedure followed is based on Rimmer and Hartmann (2014) and Zhang et al. (2013). The end members of surface and baseflow were estimated from the EC data. The surface end member is fixed at 20 μS cm^{-1}, which corresponds to the average rainfall EC as reported by Camacho Suarez *et al.* (2015) and lowest observed EC from the stream. The baseflow end member was estimated from the highest observed EC in the streamflow. There is some uncertainty in literature on the best method to define this value, so a sensitivity analysis was conducted, assuming c_b would be the percentile 90, 95, 99 or maximum (excluding outliers of the observed EC) (Kronholm and Capel, 2015). The chemical hydrograph separation using EC was then performed and a reference baseflow was estimated.

Digital filter (DF) separation using Eckhardt (2005) approach was conducted, with the α computed from recession analysis and initial estimate of BFI_{max} based on catchment geology (recommended 0.25 for perennial streams with hard rock aquifers). An optimization procedure was then used, whereby the time series of baseflow from the digital filter is compared with the tracer baseflow, goodness of fit indicators are computed, and an objective error function is minimized – i.e. the Root Mean Squared Error between reference tracer baseflow and DF baseflow. The optimal BFI_{max} parameter is the one that best fits tracer and DF methods, with minimal error, following the approach by (Rimmer and Hartmann, 2014). Both time series of baseflow were plotted for visual inspection, and goodness of fit indicators such as Nash-Sutcliffe (NS), Root Mean Squared Error (RMSE), Bias and correlation coefficient were computed. Total baseflow indices are computed and compared as well.

All the analyses were conducted at daily time steps, subsequently being aggregated to monthly and annual time scale. However, calibration was conducted only for days were EC data was available.

5.3 Results

5.3.1 Spatial and temporal variability of catchment hydrochemistry

The headwater catchments are characterized by waters with low EC (100 ± 25μS cm^{-1}, average ± standard deviation), dominated by alkalinity. The range of EC at the outlet (120 to 1200 μS cm^{-1}) is much higher than for the headwaters (40 to 400 μS cm^{-1}). The variation of EC at the outlet follows a marked seasonal pattern. EC increases during dry seasons up to 1200 μS cm^{-1} and decreases during the wet season to 200 μS cm^{-1}, reflecting the relative flow volume in each season. The variation in the Noordkaap for example, is much less accentuated with ranges between 50 to 200 μS cm^{-1} (Figure 5-3).

EC monthly variation is in the same range for the Queens (monthly average of 155 to 196 μS cm^{-1}), Noordkaap (monthly average of 98 to 123 μS cm^{-1}) and Suidkaap (monthly average of 133 to 182 μS cm^{-1}) tributaries. The EC is slightly higher in September (low flow) than in other months. Suidkaap, which is the tributary with the largest area and longest river length, exhibits more marked seasonality compared with the two smaller tributaries. The Kaap outlet shows a very strong seasonality of EC (monthly average of 405 to 711 μS cm^{-1}). There is also a pattern of higher EC in a sequence of dry years and lower values in wetter years.

The comparison of boxplots (Figure 5-4) reveals that EC, Chloride, Sodium, Calcium, Magnesium, Sulphate and Total Alkalinity are lower in the Noordkaap, followed by Suidkaap and Queens, and much higher in the Kaap; the Kaap outlet values are almost an order of magnitude higher than its contributing tributaries. For example, Chloride ranges from 1.5 to 30.3 mgL^{-1}in the Noordkaap with average 4.8 ± 2.4 mgL^{-1}, whereas the range in the Kaap outlet is 3.7 to 57.2, with average 22.9 ± 9.7 mgL^{-1}. Silica, however, has a very similar range of values across the catchments. At the Kaap outlet the mean concentration of silica (14.1± 3.0 mgL^{-1}) is only slightly higher than the tributaries (12.3 ± 2.2 mgL^{-1}), but the range of variation is much wider. Appendix A1 provides a table with detailed statistics of water quality for the Kaap catchment and its tributaries.

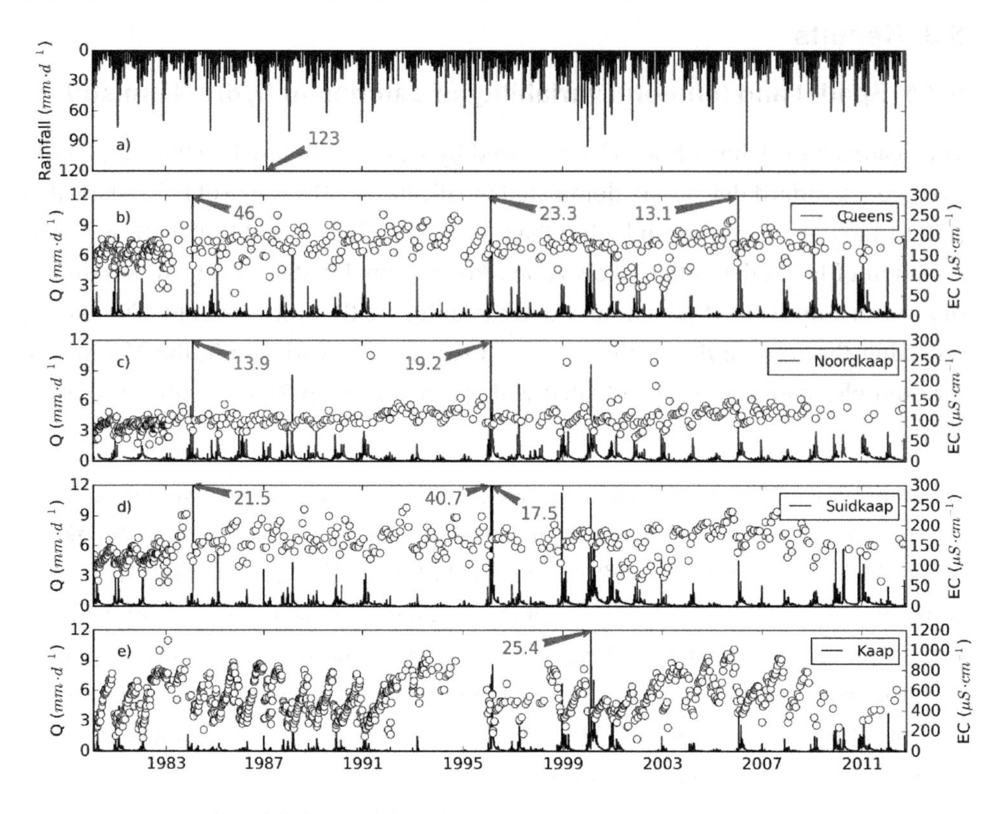

Figure 5-3. a) Areal rainfall; flow (solid black line) and EC (circles) for b) Queens, c) Noordkaap, d) Suidkaap and e) Kaap catchments; 1978-2012. Red arrows indicate rainfall and flow values higher than y-axis.

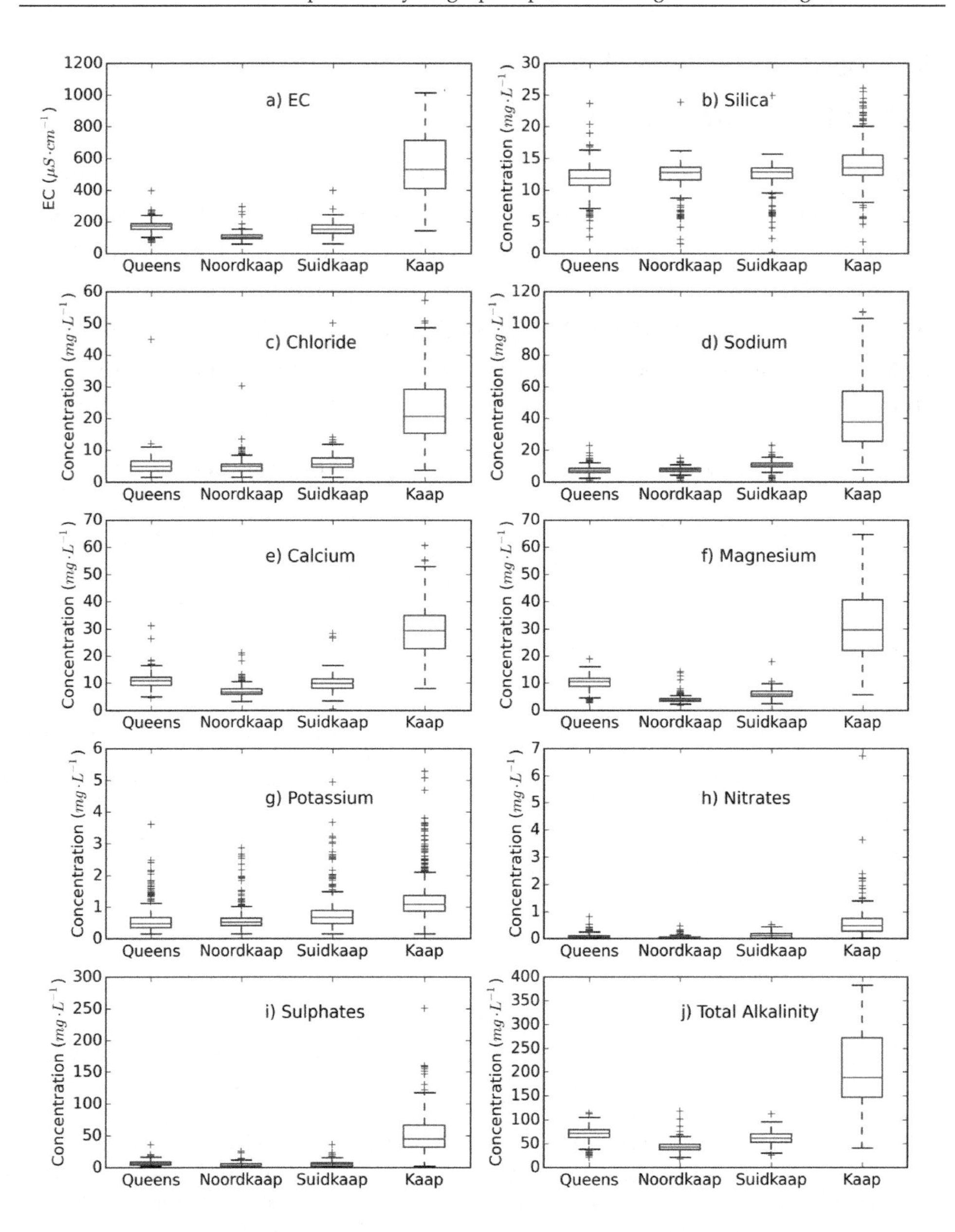

Figure 5-4. Comparison of boxplots of water quality for the four catchments: a) EC, b) Silica, c) Chloride, d) Sodium, e) Calcium, f) Magnesium, g) Potassium, h) Nitrates, i) Sulphate, j) Total Alkalinity

5.3.2 Hydrograph separation

5.3.2.1 Chemical hydrograph separation

In all hydrograph separations using EC the important contribution of baseflow to total streamflow is evident. As expected, during the dry August and September months the contribution of baseflow is relatively high, with baseflow index (BFI) of 0.82 ± 0.13 (average ± standard deviation). January and February have the lowest relative contribution with BFI 0.62 ± 0.17 in the Noodkaap tributary (Figure 5-5). In the Kaap outlet the BFI in September is 0.81 ± 0.16 whereas the BFI in February is 0.48 ± 0.16.

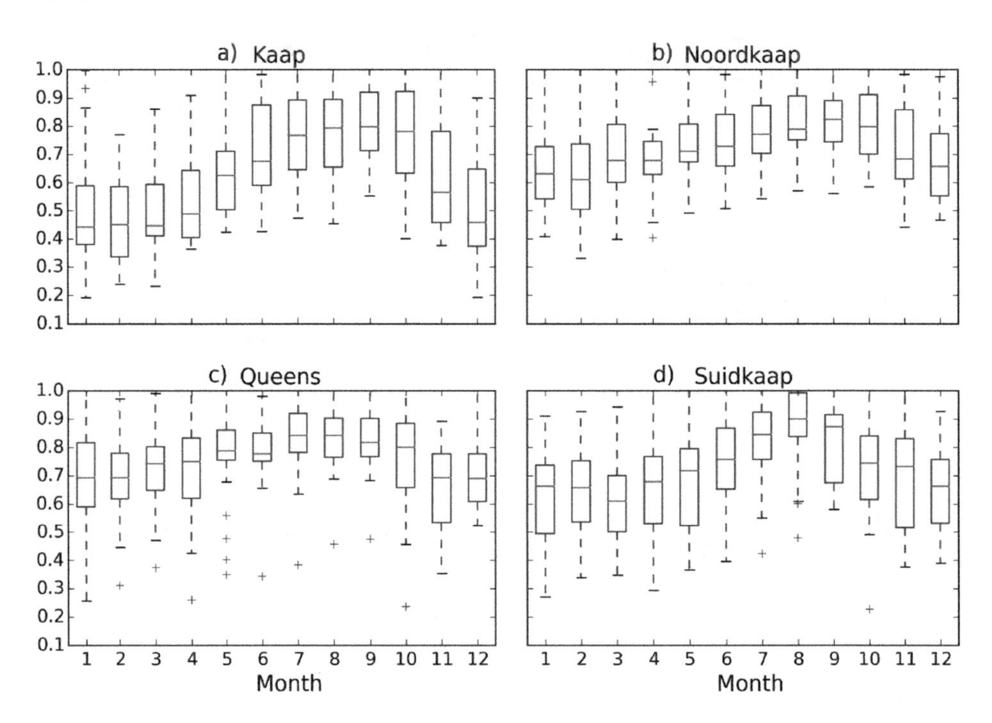

Figure 5-5. Boxplots of Baseflow Index (non-dimensional) derived from chemical hydrograph separation per month, for the period of 1978 to 2011: a) Kaap, b) Noordkaap, c) Queens, and d) Suidkaap. For this separation the groundwater end member was defined as 95% percentile of EC time series. Month 1 corresponds to January and 12 to December.

5.3.2.2 Hydrograph separation using digital filters

The hydrograph separation using digital filters was conducted within the calibration and optimization routine in Python. The parameter α was calculated from master recession curve analysis for each catchment, using three different methods (Rimmer and Hartmann, 2014). Figure 5-6 illustrates the calculation of α using the correlation method. The value of α was then fixed as the average of the 3 methods, as presented on Table 5-3.

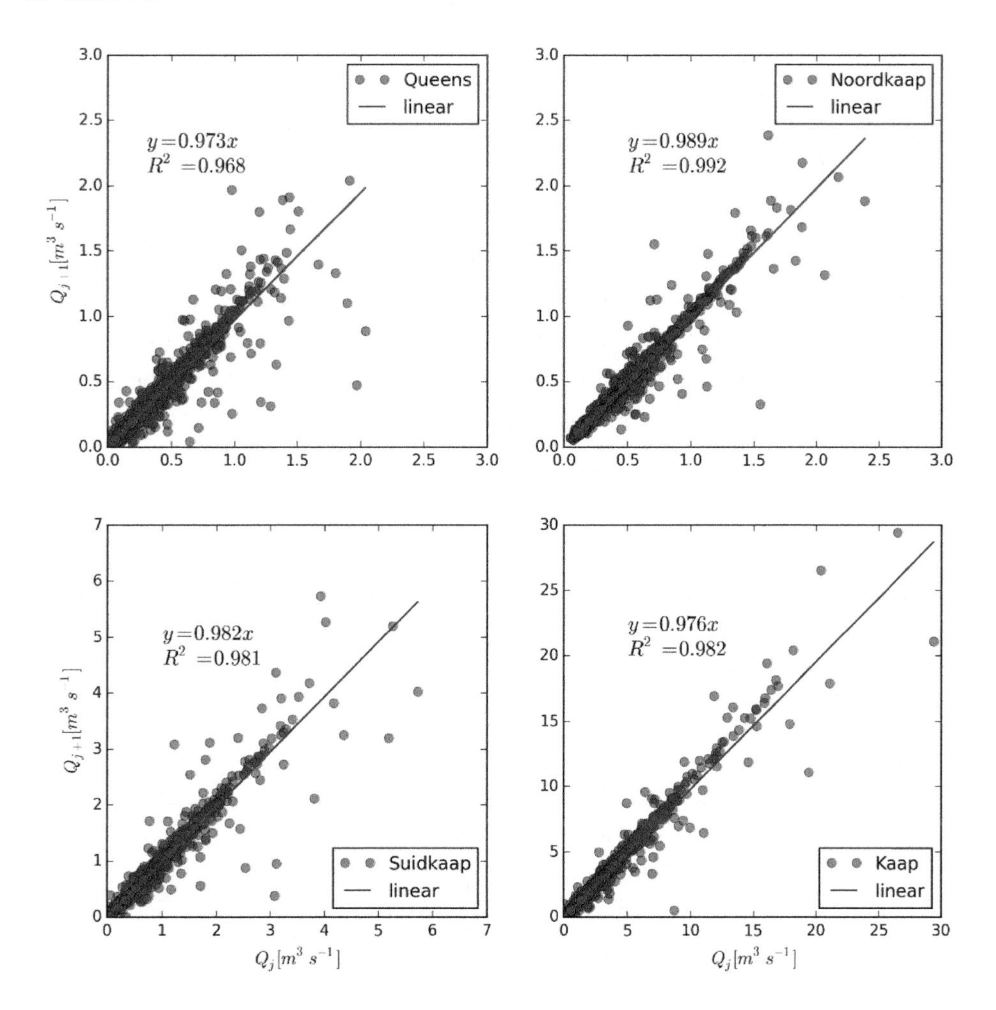

Figure 5-6 Calculation of α parameter using the correlation method (second method), for the four investigated catchments: a) Queens, b) Noordkaap, c) Suidkaap, and d) Kaap. α corresponds to the slope of the regression line through the origin in a scatterplot of streamflow Q_{j+1} against Q_j.

Table 5-3. Results of computation of parameter α using the three methods suggested by Rimmer and Hartmann (2014)

	Method 1 Master recession curve exponent	Method 2 Slope of Regression of Q_{j+1} vs Q_j	Method 3 Mean of Q_{j+1}/Q_j	Average
Queens	0.989	0.973	0.988	0.98
Noordkaap	0.995	0.989	0.993	0.99
Suidkaap	0.995	0.982	0.992	0.99
Kaap	0.985	0.976	0.983	0.98

The parameter BFI_{max} was initially estimated as BFI_{max} = 0.25 for perennial rivers dominated by hard rock aquifer (Eckhardt, 2005). This parameter was then optimized to better fit the estimated baseflow from digital filter to the reference tracer baseflow. Table 5-4 presents the results of optimized BFI_{max}, the c_s and c_b used, and the goodness of fit indicators, as well as total baseflow from tracer and digital filter separation methods.

Table 5-4. Results of calibration of digital filter parameters with chemical hydrograph separation, using EC.

Catchment	Percentile c_b	α	BFImax	c_s	c_b	Correl	RMSE	NS	BFI_tr	BFI_DF	Difference,%
Noordkaap	90	0.99	0.744	20	128	0.842	0.145	0.689	67.9	68.5	1
	95	0.99	0.652	20	138	0.816	0.140	0.628	62.5	61.7	-1
	99	0.99	0.482	20	172	0.753	0.123	0.436	48.8	47.3	-3
	Max*	0.99	0.568	20	154	0.787	0.132	0.556	55.3	55.0	-1
Suidkaap	90	0.99	0.793	20	194	0.871	0.280	0.731	64.8	66.2	2
	95	0.99	0.692	20	210	0.843	0.279	0.672	59.5	60.2	1
	99	0.99	0.637	20	236	0.828	0.275	0.647	52.5	56.7	8
	Max*	0.99	0.539	20	246	0.787	0.267	0.540	50.2	50.1	0
Queens	90	0.98	0.964	20	205	0.961	0.236	0.910	69.8	84.2	21
	95	0.98	0.955	20	222	0.950	0.256	0.881	64.0	82.0	28
	99	0.98	0.935	20	255	0.925	0.290	0.817	55.1	78.0	42
	Max*	0.98	0.942	20	242	0.934	0.278	0.842	58.3	79.4	36
Kaap outlet	90	0.98	0.500	20	784	0.821	0.133	0.653	44.3	44.8	1
	95	0.98	0.412	20	833	0.791	0.128	0.556	41.7	38.3	-8
	99	0.98	0.341	20	921	0.773	0.118	0.468	37.6	32.6	-13
	Max*	0.98	0.271	20	1100	0.757	0.101	0.375	31.4	26.3	-16

*Maximum EC excluding outliers

Note.c_b is the EC concentration of groundwater end member ($\mu S\ cm^{-1}$), c_s is the EC surface water end member ($\mu S\ cm^{-1}$), α and BFI_{max} are digital filter parameters, Correl, RMSE and NS are indicators of goodness of fit. BFI_Tr is total baseflow (%) using chemical hydrograph separation BFI_DF is total baseflow (%) using digital filter, and their difference.

5.3.3 Calibration of digital filter using tracers and sensitivity analysis

Figure 5-7 shows the results of the calibration exercise (conducted for the entire time series) for the four catchments for the period 1997 to 2005.

The digital filter fills the gaps where tracer data is not available, and it is possible that it overestimates the peaks, given that less tracer data is available during peaks compared to normal or low flow. In some occasions, the digital filter also seems to underestimate or overestimate the low flows, particularly at Suidkaap and Queens. However, it is evident that the variability of the hydrograph is generally well captured by the calibrated digital filter results.

Overall, good to very good correlations were achieved between reference tracer separation and the calibrated digital filter. Correlation coefficient ranged from 0.753 to 0.961 for all the calibrations. The Nash-Sutcliffe (NS) coefficient ranged from 0.375 to 0.91 with RMSE ranging from 0.10 to 0.29 (Table 5-4).

The optimal BFI_{max} for the Noordkaap catchment ranged from 0.482 (99% used for c_b) to 0.744 (90% for c_b), which corresponded to total baseflow of 48 to 68% respectively. The correlation between digital filter baseflow and reference tracer baseflow ranged from 0.753 to 0.842, which shows a good correlation between them. The difference between total baseflow from calibrated DF and tracer was minimal. The RMSE ranged from 0.123 to 0.145.

The Queens catchment however showed some distinct results. The tracer separation suggested high baseflow contributions, resulting in even higher baseflow contributions for the calibrated DF separation. Optimal BFI_{max} ranged from 0.935 to 0.964, with very high correlation coefficients (0.925 to 0.961). Using the tracer method total baseflow was 55 to 70%, while for the calibrated filter it was 78 to 84%. This resulted in a difference of 21 to 42%. The RMSE ranged from 0.236 to 0.29, whereas the NS varied from 0.817 to 0.91.

The Kaap outlet had the lowest BFI_{max}, ranging from 0.271 to 0.50, which yielded a total baseflow of 26 to 45% using the calibrated filter. RMSE were the lowest, ranging from 0.10 to 0.133, whereas NS were also the lowest (0.375 to 0.653).

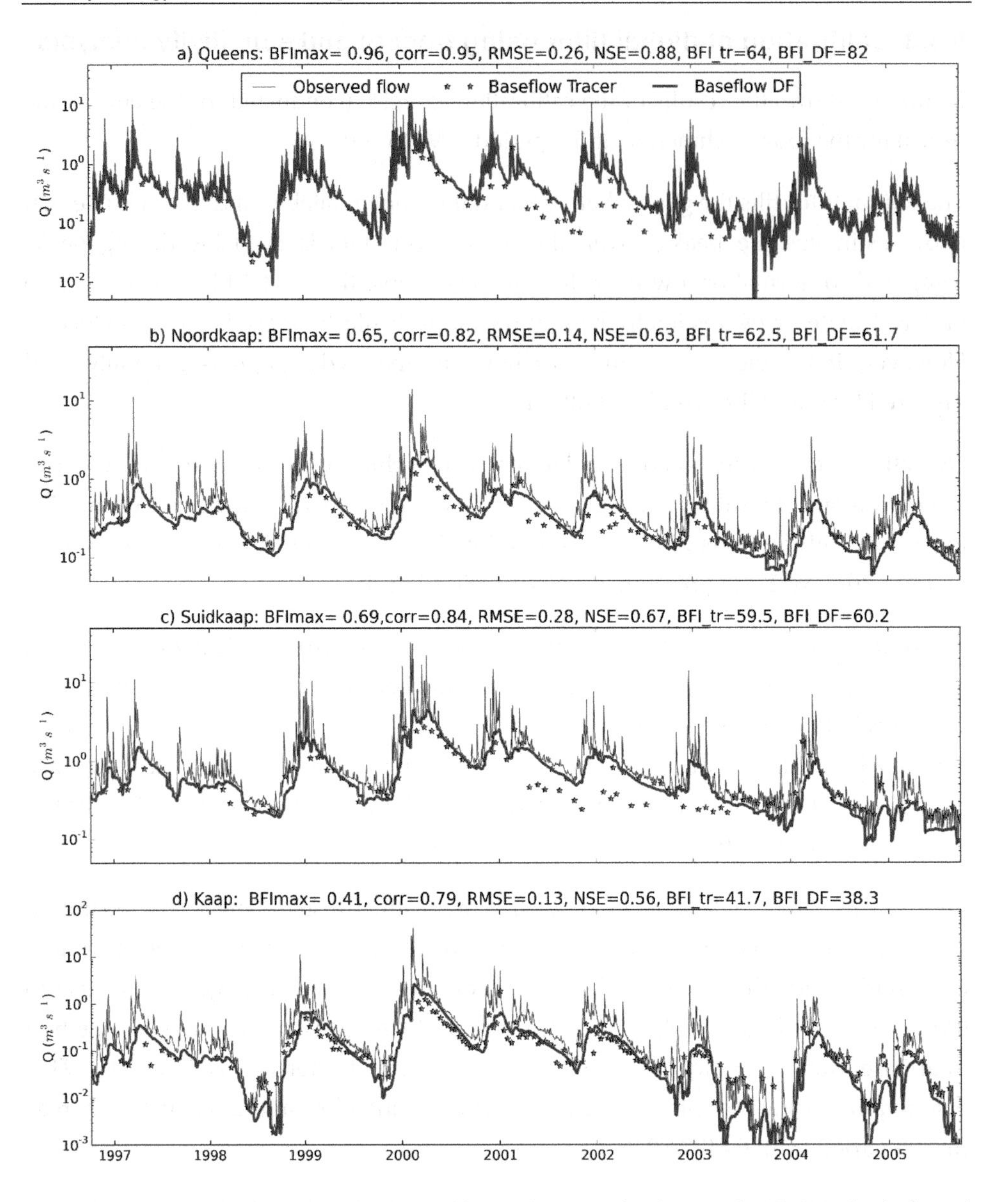

Figure 5-7 Results of calibration of the digital filter with tracers, for the period 1997-2005 (hydrological years) for the four investigated catchments: a) Queens, b) Noordkaap, c) Suidkaap, and d) Kaap. Baseflow DF is the baseflow resulting from digital filter separation and Baseflow Tracer is the baseflow estimated from chemical hydrograph separation.

5.3.4 Runoff components

5.3.4.1 Monthly scale

The results of digital filter hydrograph separation with calibrated parameters using cb of 95% percentile were aggregated to monthly volumes. Figure 5-8 illustrates the average monthly flow components for the catchments, and Appendix A2 presents more detailed results of both monthly and annual flow components. The variability of flow components in a monthly scale is high, due to the high variability of rainfall and thus flow. Highest baseflow and quickflow volume contributions occur in February and March.

In the Queens catchment, the baseflow component ranges from $0.32 \cdot 10^6 \pm 0.3 \cdot 10^6$ m^3 (average ± standard deviation) in August to $2.28 \cdot 10^6 \pm 2.61 \cdot 10^6$ m^3 in February. The average monthly quickflow is highest in January ($0.76 \cdot 10^6 \pm 1.2 \cdot 10^6$ m^3) and lowest in August ($0.02 \cdot 10^6 \pm 0.02 \cdot 10^6$ m^3). The highest baseflow ($11.5 \cdot 10^6$ m^3) occurred in the flood of February 1996. Highest quickflow ($6.2 \cdot 10^6$ m^3) however occurred during the flood of January of 1984.

The Noordkaap catchment exhibits a similar pattern, but with highest average monthly baseflow in March ($1.47 \cdot 10^6 \pm 0.99 \cdot 10^6$ m^3) and lowest in October ($0.44 \cdot 10^6 \pm 0.2 \cdot 10^6$ m^3). The quickflow ranges from $0.13 \cdot 10^6 \pm 0.07 \cdot 10^6$ m^3 in June to $1.24 \cdot 10^6 \pm 1.65 \cdot 10^6$ m^3 in February. In this catchment the peak baseflow occurred in March 1996 ($4.53 \cdot 10^6$ m^3) and highest quickflow during the flood of February 2000 ($7.4 \cdot 10^6$ m^3).

In the Suidkaap catchment, the baseflow component ranges from $0.69 \cdot 10^6 \pm 0.49 \cdot 10^6$ m^3 in October to $2.28 \cdot 10^6 \pm 2.1 \cdot 10^6$ m^3 in March. The average monthly quickflow is highest in February ($2.1 \cdot 10^6 \pm 3.88 \cdot 10^6$ m^3) and lowest in June ($0.22 \cdot 10^6 \pm 0.13 \cdot 10^6$ m^3). Similar to the other catchments, the peaks of baseflow ($10.2 \cdot 10^6$ m^3) and quickflow ($10.2 \cdot 10^6$ m^3) occurred during the floods of 2000 and 1996.

In the Kaap catchment, the monthly average baseflow component ranges from $0.77 \cdot 10^6 \pm 0.87 \cdot 10^6$ m^3 in October to $7.34 \cdot 10^6 \pm 11.33 \cdot 10^6$ m^3 in March. The average quickflow is highest in February ($16.1 \cdot 10^6 \pm 37.0 \cdot 10^6$ m^3) and lowest in August ($0.84 \cdot 10^6 \pm 0.91 \cdot 10^6$ m^3). For the Kaap the peak baseflow occurred in March 2000 ($54.5 \cdot 10^6$ m^3) and highest quickflow during the flood of February 2000 ($193.7 \cdot 10^6$ m^3). In this catchment there are several months with minimum flow, baseflow and quickflow of zero, occurring during the drought of 1992-94.

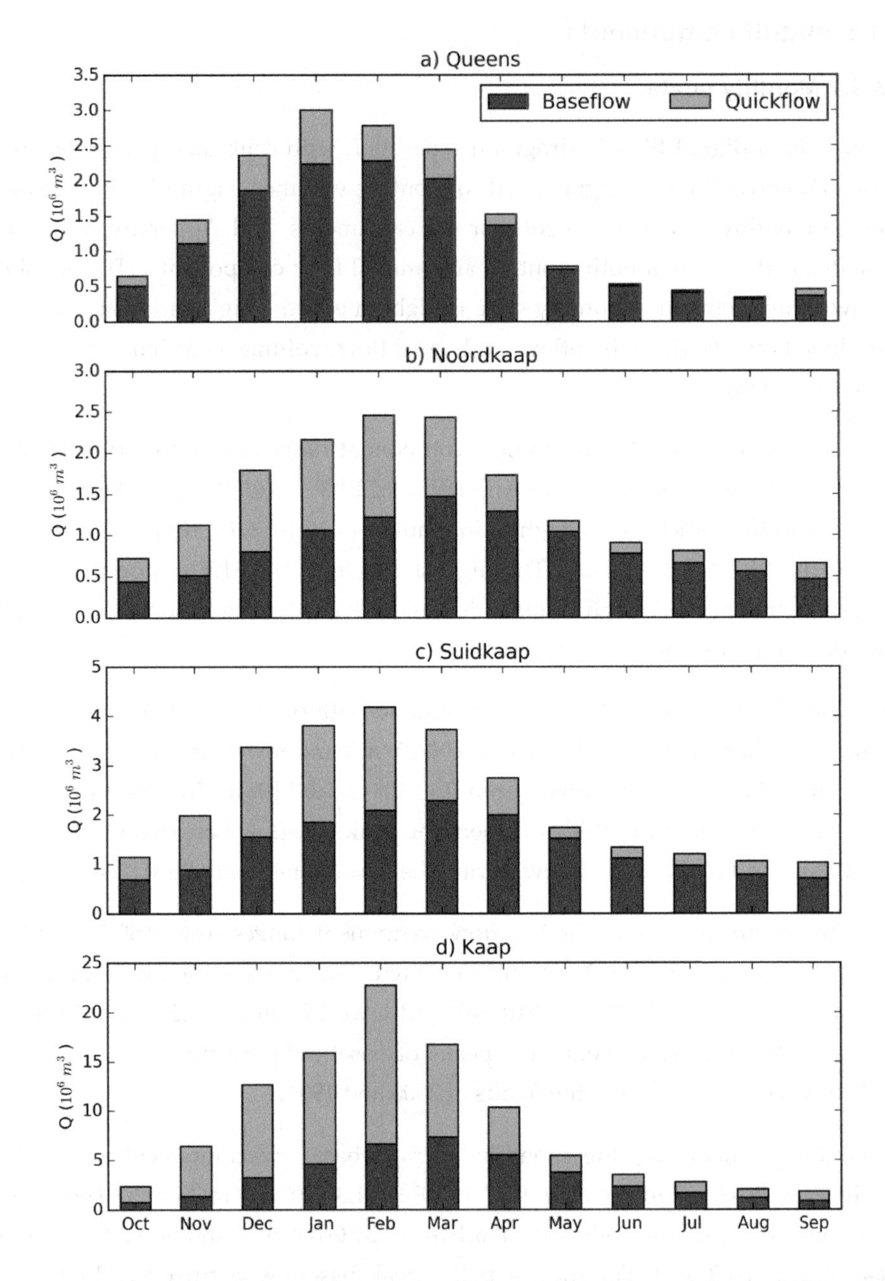

Figure 5-8 Monthly average flow components using calibrated digital filters: a) Queens, b) Noordkaap, c) Suidkaap, and d) Kaap

5.3.4.2 Annual scale

In Figure 5-9, the results of digital filter separations are aggregated to hydrological year total volumes for the entire analysis period (1978-2012). At the annual scale, it is evident that in drier years most of the flow is composed of baseflow, whereas in wetter years the baseflow contribution is close to the BFI_{max} percentage (40 to 70%).

In the Queens catchment, the baseflow contribution ranges from $1.4 \cdot 10^6$ to $42.3 \cdot 10^6$ m^3 y^{-1}. The lowest contribution occurred during the drought of 1992-94 and in 1983, whereas the highest occurred in 2000 and 2011. The annual average baseflow is about $13.7 \cdot 10^6$ m^3 y^{-1}, which corresponds to about 82% of the total flow. The quickflow ranges from $0.4 \cdot 10^6$ to $8.1 \cdot 10^6$ m^3 y^{-1}, with average of $3.0 \cdot 10^6$ m^3 y^{-1}, corresponding to a 0.23 ratio of quickflow to baseflow.

The Noordkaap has a lower range of baseflow, varying from $2.9 \cdot 10^6$ to $24.5 \cdot 10^6$ m^3 y^{-1}. The annual average baseflow is $10.3 \cdot 10^6$ m^3 y^{-1}, constituting 62% of total flow. The quickflow ranges from $2.0 \cdot 10^6$ to $15.7 \cdot 10^6$ m^3 y^{-1}, with average $6.4 \cdot 10^6$ m^3 y^{-1}, corresponding to 0.62 quickflow to baseflow ratio.

The total baseflow contribution in the Suidkaap is about 60% of total flow, with average baseflow of $16.4 \cdot 10^6$ m^3 y^{-1}. The average quickflow contribution is $10.9 \cdot 10^6$ m^3 y^{-1}, which corresponds to 0.66 ratio of quickflow to baseflow. During the analysis period (hydrological years of 1978 to 2012), the baseflow ranged from $3.2 \cdot 10^6$ to 56.7 $\cdot 10^6$ m^3 y^{-1} and quickflow from $2.4 \cdot 10^6$ to $38.0 \cdot 10^6$ m^3 y^{-1}.

At the Kaap outlet the range of variation of baseflow is the widest, with a minimum of $0.3 \cdot 10^6$ to a maximum of $237.9 \cdot 10^6$ m^3 y^{-1}. The Kaap also had the highest contributions from quickflow, ranging from $1.2 \cdot 10^6$ to $358.5 \cdot 10^6$ m^3 y^{-1}, with average $63.4 \cdot 10^6$ m^3 y^{-1}. Similar to the other catchments, the minimum baseflow occurred in the drought of 1994 and the maximum in the wet year 2000. The average baseflow at the outlet is $39.4 \cdot 10^6$ m^3 y^{-1} and total baseflow contribution is about 38% of total flow. The quickflow to baseflow ratio is 1.61, which is the highest of all catchments, highlighting the importance of quickflow in the Kaap River valley.

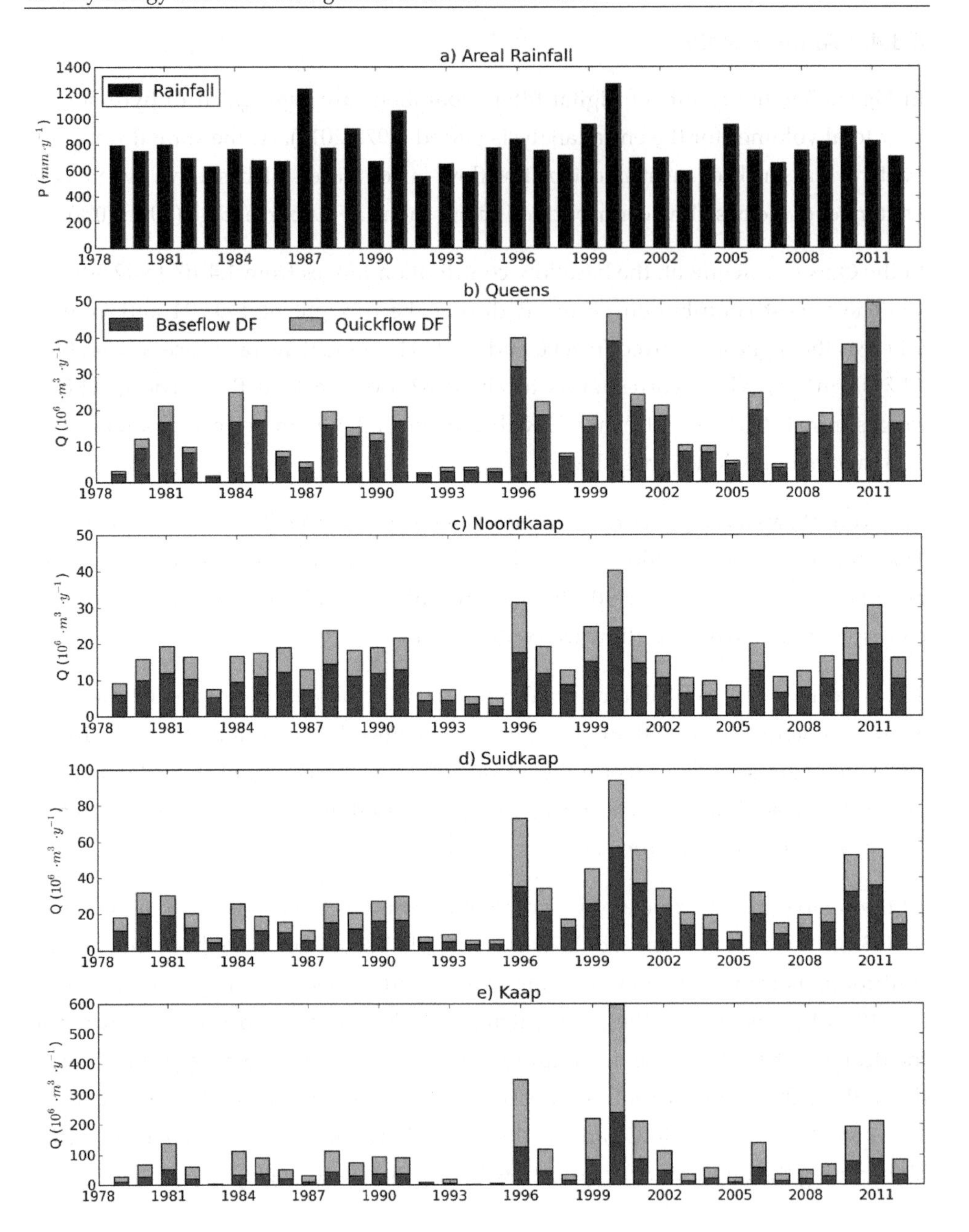

Figure 5-9 Summary of Hydrograph separation using calibrated digital filters, 1978-2012: a) Areal Rainfall, b) Queens, c) Noordkaap, d) Suidkaap, and e) Kaap. The rainfall and hydrograph is aggregated by hydrological year total volumes. Note the different vertical scales.

5.4 Discussion

5.4.1 Hydrochemical analysis and applicability of tracers for hydrograph separation

The analysis of water quality data revealed that EC was a suitable tracer to perform hydrograph separation for the Kaap catchment. This is in line with several authors who have used high frequency and discrete time series of EC to perform hydrograph separation and quantify baseflow at a daily time scale (Zhang *et al.*, 2013; Miller *et al.*, 2014; Miller *et al.*, 2015; Longobardi *et al.*, 2016). For example, Miller *et al.* (2015) used discrete EC data sets to quantify baseflow contributions in the Upper Colorado basin, using hydrograph separation. They were able to quantify this component of the water budget, in a very heavily regulated and snowmelt dominated catchment. Hughes *et al.* (2003) used recursive digital filters to derive baseflow at monthly time scale for catchments in South Africa. They were able to derive baseflow parameters for use in the Pitman model, which is widely used for water resources assessment in South Africa. Li *et al.* (2014) estimated baseflow and groundwater recharge based on a similar approach of EC and recursive digital filter for a small watershed in Canada.

The high variability of water quality parameters including EC constitutes an important challenge to the end member definition of chemical hydrograph separation, as already noted by many authors (Miller *et al.*, 2014). This makes the quantification of runoff components at various time scales rather indicative than exact, however, the temporal course and behaviour reveals relevant insights into the dynamics of the system.

EC is strongly correlated with other hydrochemical parameters, such as Calcium, Magnesium, and Chloride (see annex A1 for multivariate analysis of water quality parameters). Most hydrochemical parameters exhibit a pattern of dilution during the wet season (October to March) and concentration in the dry season (April to September); this is likely due to high evaporation rates, lower flow available during the dry season, and relative stronger contributions from groundwater sources and possible anthropogenic sources (irrigation return flow and wastewater). Some studies in South Africa, for example Sahula (2014), also report that increased urbanization and informal settlements contribute with both point and non-point pollution, due to discharges from untreated sewage. In the Kaap catchment, this is combined with abandoned mines and irrigation return flows thereby impacting the water quality in the river significantly (Deksissa et al., 2003; Slaughter and Hughes, 2013; Retief, 2014; Sahula, 2014).

EC increases dramatically after the confluence of the Noordkaap and Suidkaap River. This is also the section where there is a major change in geology, topography and land use. Furthermore, the land use includes some industries, mining and irrigated agriculture, which will likely increase the load that contributes to EC through effluents and return flows (Deksissa *et al.*, 2003; Retief, 2014). The input of return flows were not taken into account for the two-component hydrograph separation, given that not enough information is available regarding the timing, volume and load of these. Further research is required to quantitatively assess which factor could be the most important cause of this EC increase.

5.4.2 Derivation and validity of results obtained by digital filters

Digital filters are very useful and inexpensive to quantify baseflow contributions. Recursive digital filters in particular are the most developed method to perform hydrograph separation using the streamflow records alone (Eckhardt, 2008).

However, given the high sensitivity of the digital filter parameters, it is important that calibration of parameters is done prior to operational use of baseflow estimates. When water quality data is available through routine sampling campaigns, or dedicated experiments, it can be used to calibrate digital filters. It is recommended that high frequency water quality data be used in a number of sites in the catchment to perform the calibration of digital filters (Zhang *et al.*, 2013; Longobardi *et al.*, 2016). Once the filter parameters are calibrated, the entire flow time series can be used to derive a similar time series of baseflow, such as in the study of Zhang *et al.* (2013).

The parameter α, for example, can be determined by recession curve analysis (Table 5-3). The parameter BFI_{max} requires more calibration, as it is more sensitive than α, and cannot be physically measured (Eckhardt, 2005; Eckhardt, 2012; Rimmer and Hartmann, 2014). In this study, we found that once α is determined, BFI_{max} can be optimized to fit observed time series of EC. BFI_{max} is more sensitive to the baseflow end member of EC. Thus, for the sensitivity analysis we looked at 90, 95 and 99% percentiles of EC for c_b definition. For 90 and 95, the results are quite similar and comparable, but when 99% c_b is used, the BFI_{max} is much smaller than with other percentiles. This finding is critical, because the 99% c_b could include some EC outliers that greatly bias the separation. Therefore, the 95% was adopted for this study as reference end member for c_b.

This study demonstrates that data collected by routine campaigns (weekly/monthly) can be useful for calibration of digital filters, similar to the studies of Rimmer and

Hartmann (2014) and Li *et al.* (2014). The calibration data can be discrete but with greater temporal resolution this improves, as the definition of end members is based on a more informative data set (covering a range of flow/EC relationships). Additionally, we found that the calibrated BFI_{max} is, in most cases, higher than the BFI_{max} values found in the literature for perennial rivers with predominant hard rock aquifers (Eckhardt, 2005).

5.4.3 Implications for hydrological process understanding

This study revealed the importance of seasonality on the flow generation in the Kaap catchment. Baseflow is a very important component of river flow, particularly during the dry season (May-September). During the wet season (October – March), baseflow still contributes about 40 to 60% of river flow. Hughes *et al.* (2003) also reported very high baseflow contribution of over 60% in a tributary of the nearby Sabie River.

In the Kaap River, however, the volumetric baseflow contribution appears to be lower than in other sub-catchments. Camacho Suarez *et al.* (2015) conducted an intense tracer study in the Kaap catchment during the wet season of 2013/2014. They used both hydrochemical and environmental tracers, and reported groundwater contribution of 64 to 98% in four events studied, based on daily and shorter time scale data. This seems to contradict the digital filter results, but a lot of water abstractions (mainly surface water) occur on the lower Kaap valley. DWAF (2009b) report a crop water requirement of $91.7 \cdot 10^6$ m^3 y^{-1}, for about 98 km^2 of sugarcane irrigation. They further report that $62 \cdot 10^6$ m^3 y^{-1} is supplied to the irrigation boards– this is about 33% of the Naturalized Mean Annual Runoff of $189 \cdot 10^6$ m^3 y^{-1}(Bailey and Pitman, 2015). Most of the water is abstracted from the river flow during the dry months, which explains the apparent low baseflow contributions in the Kaap catchment. In the wet season, the water demand for irrigation is reduced; therefore the quickflows are not greatly affected.

Furthermore, Chapter 3 studied the drivers of streamflow variability in the Incomati Basin (where the Kaap catchment is a tertiary sub-catchment). It was found that plantations (eucalyptus and pine) and irrigation caused a significant decline on streamflow in several streamflow gauges in the Crocodile catchment, including the Noordkaap and Kaap. About $39.8 \cdot 10^6$ m^3 y^{-1} is attributed to afforestation in the Kaap catchment (DWAF, 2009b), which is considered a streamflow reduction activity in South Africa. This finding was also supported by van Eekelen *et al.* (2015), in a study were remote sensing was used to estimate direct and indirect water withdrawals in the Incomati basin.

5.4.4 Application in water resources management

The flow components information is important for the understanding and quantification of environmental flows, as well as to better parameterize the hydrological models used for water resources planning and management, particularly rainfall-runoff models (O'Brien *et al.*, 2014).

Another finding of this study is that the simple model of two component mixing can work relatively well at smaller headwater catchments (such as the Noordkaap in this study), but it is not sufficiently complex to characterize the runoff generation processes that occur at a meso-scale catchment (such as the Kaap). This can also be explained by the fact that the headwater catchments often have a more homogenous geology and land use, but in a larger catchment the heterogeneity of geology, soils and land use increases as well as human influence can introduce new components, for instance, return flows from irrigated agriculture. For the larger catchment more intense monitoring is required to account for different flow contributions and anthropogenic activities.

This research further highlights that even though the chemical hydrograph separation is a useful concept, it is quite difficult to determine the exact values of end member components at meso-scale catchments. In fact, these end members are not physically observed and measured in one spot but are the result of an assumed mixing of various sources in a complex reality. The rainfall input, for example, is distributed over a large catchment area and interacts with the land surface, soils and stagnant water before it reaches the river and becomes streamflow. The amount of rainfall that falls directly on the river course is insignificant compared to the amount that is routed as surface runoff during intense rainfall events. Therefore, even though we assume that this component resembles rainfall water quality, it is likely to have a much more diverse water quality signature.

The baseflow, on the other hand, is also composed of components from different sources. Baseflow, which is the flow that feeds the river during extended periods without rainfall, can be composed of deep regional groundwater discharge, localized shallow groundwater, return flows from irrigated agriculture, discharge from municipal water use, discharge from informal settlements, or releases from reservoirs, among others.

The use of calibrated digital filters still reveals important insights of the catchment flow dynamics at daily, monthly and annual scales, which are useful for quantification of environmental flows, river operation and conjunctive groundwater management.

5.5 Conclusions

In this study, we analysed historical hydrochemical data and used it to perform two-component hydrograph separations at monthly and annual time scale in a meso-scale semi-arid catchment and three sub-catchments. There is strong seasonality in the runoff and hydrochemistry in all catchments, particularly at the main outlet.

Electrical conductivity was identified as a suitable tracer to perform chemical hydrograph separation for these catchments, given the consistency of the data set. The chemical hydrograph separation indicates that the baseflow dominates the total flow, with contributions ranging from 50% in wet season to 90% in dry season.

Hydrograph separation was also performed using Eckhardt's recursive digital filter, with daily streamflow data. The parameter BFI_{max} was calibrated for different sets of groundwater end members, using the chemical hydrograph separation for reference.

The digital filter parameters are very sensitive, and their use without calibration is not recommended, as they can yield very different quantitative results. Optimal sets of α and BFI_{max} were identified for the studied catchments. In spite of the uncertainties in α and BFI_{max}, the digital filter hydrograph separation is very useful to interpolate and to extend baseflow estimates for periods where tracer data is not available. It is recommended that more calibration studies are conducted in semi-arid catchments to assess if regionalization (transfer in space) of filter parameters is possible.

Another important finding of this study is the high contribution of baseflow to total flow during both wet and dry conditions. This means that the groundwater reservoirs respond quickly during storm events, which is important to consider for flood forecasting, environmental flow assessments, and for land use planning and management, in order to optimize/enhance groundwater recharge or prevent practices that compromise this.

This study has tested the usefulness of using readily available secondary water quality data to calibrate hydrograph separation using recursive digital filters. We also tested a method to optimize the BFI_{max} parameter used in digital filter methods for hydrograph separation. Once the parameters for the digital filter separation are calibrated, it is possible to perform hydrograph separation for much more extensive periods of time than the available water quality data. Also, the digital filter performs separation in a daily basis, whereas water quality data is only available at weekly or monthly time steps. So a more refined separation is possible using this approach.

The relevance of this analysis is that it allows for estimation of the baseflow component on a daily basis, from readily available streamflow and water quality data. Thus, these findings can be used to improve rainfall-runoff models, especially in terms of conjunctive groundwater management, river operations, and quantification of environmental flows where decisions regarding releases from dams and/or abstractions from rivers are done on a daily/weekly basis.

5.6 Supporting documentation

Appendix A1 - Table with detailed statistics of water quality for the Kaap catchment and its tributaries. Descriptive statistics and correlation matrix of the water quality parameters, which include: Electrical conductivity, pH, Calcium, Magnesium, Potassium, Sodium, Chloride, Sulphates, Total Alkalinity, Silica, Fluoride, Nitrates, Ammonia, Phosphate and Total Dissolved Salts

Appendix A2 - Results of calibrated monthly and annual flow components for Kaap, Suidkaap, Noordkaap and Queens Catchments

6

UNDERSTANDING RUNOFF PROCESSES IN A SEMI-ARID ENVIRONMENT THROUGH ISOTOPE AND HYDROCHEMICAL HYDROGRAPH SEPARATIONS

In this chapter the understanding of runoff generation processes is further enhanced through process studies. An intense sampling campaign was conducted in the Kaap catchment in the wet season of 2013/2014. Through a hydrochemical characterization of surface water and groundwater sources of the catchment and two and three component hydrograph separations, runoff components of the Kaap catchment were quantified using both hydrochemical and isotope tracers. End-member mixing analysis allowed the identification of the relevant runoff components. Hydrograph separation results showed that runoff in the Kaap catchment is mainly generated by groundwater sources. Relationships between rainfall and runoff were also explored, to further understand the runoff generation mechanisms. Strong correlations were found between antecedent precipitation and direct runoff. Finally, the complexity of runoff processes understanding in the context of semi-arid areas is also discussed.

This chapter is based on: Camacho Suarez VV, Saraiva Okello AML, Wenninger JW, Uhlenbrook S. 2015. Understanding runoff processes in a semi-arid environment through isotope and hydrochemical hydrograph separations. *Hydrol. Earth Syst. Sci.*, 19: 4183-4199. DOI: 10.5194/hess-19-4183-2015.

6.1 Introduction

Understanding runoff processes facilitates the evaluation of surface water and groundwater risks with respect to quality and quantity (Uhlenbrook *et al.*, 2002). It assists in quantifying water resources for water allocation, hydropower production, design of hydraulic structures, environmental flows, drought and flood management, and water quality purposes (Blöschl *et al.*, 2013a). The need for understanding runoff processes has led to the development of tools such as hydrograph separation techniques that identify runoff components in stream water, flowpaths, residence times and contributions to total runoff (Weiler *et al.*, 2003; Hrachowitz *et al.*, 2009; Klaus and McDonnell, 2013). Several hydrograph separation studies using environmental isotopes and geochemical tracers have been carried out in forested, semi-humid environments which have led to new insights of runoff processes in these areas (e.g. Pearce *et al.*, 1986; Bazemore *et al.*, 1994; Burns *et al.*, 2001; Uhlenbrook *et al.*, 2002; Tetzlaff and Soulsby, 2008). But, there is still a need for understanding runoff generation mechanisms in tropical, arid and semi-arid areas as they were much less investigated (Burns, 2002).

Studying runoff processes in arid and semi-arid regions may be a challenging task due to the high temporal and spatial variability of rainfall, high evaporation rates, deep groundwater resources, poorly developed soils, and in some cases the lack of surface runoff (Wheater *et al.*, 2008; Hrachowitz *et al.*, 2011; Blöschl *et al.*, 2013a). Although these challenges may not be applicable in various instances (e.g. a reduced vegetation cover results in less importance of the interception process), arid and semi-arid regions may face extra difficulties due to the remoteness of some of these areas and financial constraints, as many of them are located in developing countries.

Arid and semi-arid regions are characterized by its sporadic, high-energy, and low frequency precipitation occurrence (Wheater *et al.*, 2008; Camarasa-Belmonte and Soriano, 2014). Dry spells can last for years, and rain events may vary from a few millimeters to hundreds of millimeters per year. High intensity storms may generate most if not all the season's runoff (Love *et al.*, 2010a). These events can also increase erosion, reduce soil infiltration capacity and enhance the surface runoff (Camarasa-Belmonte and Soriano, 2014). On the contrary, the lack of precipitation may result in reduced to non-existent groundwater recharge. Compared to humid regions, where evaporation is generally limited by the amount of energy available, evaporation in arid and semi-arid areas is usually limited by the water availability in the catchment (Wang *et al.*, 2013). Evaporation becomes the dominant factor in driving the hydrology of arid and semi-arid areas. Understanding the impact of evaporation on stream runoff processes becomes more complex due to the spatial variability of vegetation. An increase in vegetation cover due to a wetter rainfall season may result in higher evaporation rates, reduced streamflow and an increase in soil infiltration capacity (Mostert *et al.*, 1993; Hughes *et al.*, 2007). Transmission losses through the stream channel bed may also reduce the total runoff and increase the volume of recharged groundwater. This occurrence is evident in the overall Incomati Basin, where downstream areas (e.g. Mozambique) benefit from transmission losses and return flows of upstream areas (Nkomo and van der Zaag, 2004; Sengo *et al.*, 2005).

This paper explores the runoff processes, including surface-groundwater interactions in the Kaap Catchment, South Africa by describing the spatial hydrochemical characterization of the catchment, separating the runoff components through isotope and geochemical tracer analysis, and determining the suitability of isotopic tracers for the characterization of runoff components in the catchment.

6.2 Study area

The Kaap catchment is described in Section 5.2.1. Figure 6-1 shows the location of meteorological stations used, and Figure 6-2 presents key catchment characteristics relevant for this chapter.

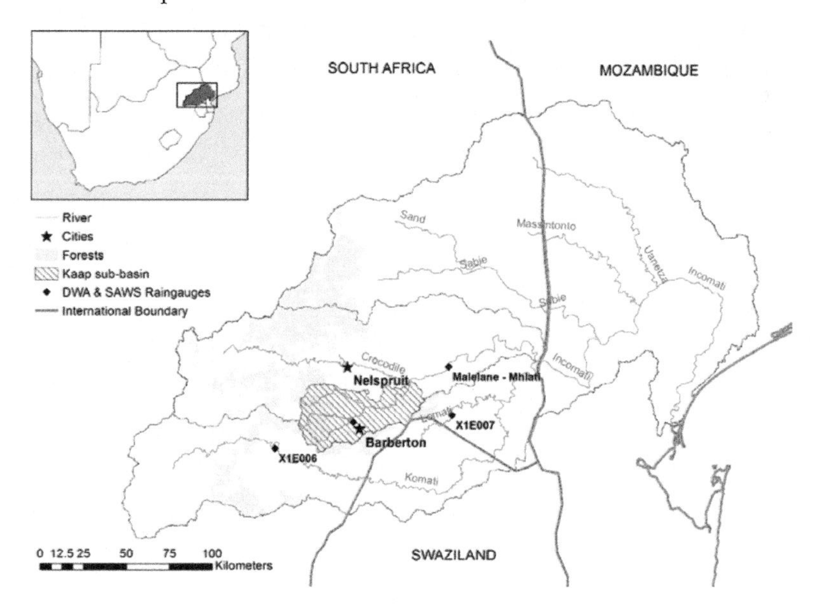

Figure 6-1. Location of the Kaap catchment in the Incomati basin displaying nearby cities, and DWA and SAWS rain gauges

Class-A-pan evaporation rates largely exceed precipitation during most parts of the year. The range of long- term potential evaporation (PET) shown in Figure 6-2F for the catchment (1950-2000) is between 1500 to 1900 mma^{-1} (Schulze, 1997). The PET data show that most of the catchment is semi-arid, according to UNEP definition (UNEP, 1997), as illustrated on Figure 6-2E (Aridity Index = Mean Annual Precipitation / Mean Annual Potential Evaporation). However, according to the Köppen-Geiger classification the Kaap catchment is sub-tropical.

Although analytical methods for hydrograph separation have been carried out in the Kaap River, no accurate estimations of runoff components were retrieved in the area. Thus, this paper also provides a baseline for understanding surface and groundwater dynamics in the Incomati trans-boundary River system. The Kaap River is a major contributor of flow to the Crocodile River which flows into the Incomati trans-boundary River. The Incomati waters are shared by South Africa, Swaziland and Mozambique, where the need to avoid tensions related to the management of water resources have led to the development of water-sharing agreements such as the Tripartite Interim Agreement on Water Sharing of the Maputo and Incomati Rivers (Van der Zaag and Carmo Vaz, 2003). The need for reliable data and understanding of the hydrological functioning of the system has been highlighted in these agreements (Slinger *et al.*, 2010). In addition, the Kaap River and the neighboring catchments have experienced devastating floods in February 2000 and March 2014 with return periods exceeding 200 years (Smithers *et al.*, 2001) .

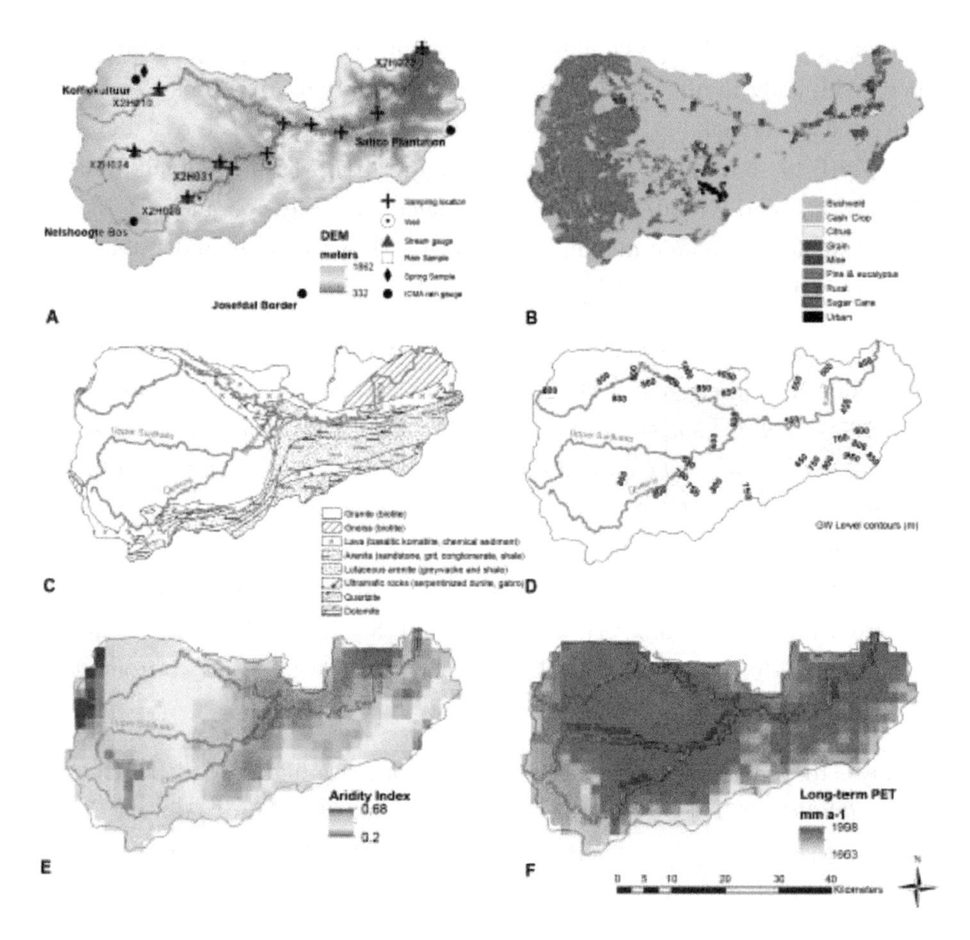

Figure 6-2. (a) Digital elevation model (DEM) of the Kaap catchment with sampling locations and stream and rain gauge locations, (b) land use map, (c) geological map, (d) piezometric map of static groundwater levels, (e) aridity index (<0.03 hyper arid, 0.03–0.2 arid, 0.2–0.5 semi-arid, 0.5–0.65 dry sub-humid, >0.65 humid) and (f) long-term mean potential evapotranspiration (PET). GIS layers are courtesy of the Water Research Commission (2005), South Africa.

6.3 Data and methods

6.3.1 Long-term datasets

Hydrological data in the catchment, including precipitation, evaporation, streamflow and groundwater records, were collected from the Department of Water and Sanitation (DWS), the South African Weather Service (SAWS), the South African Sugarcane Research Institute (SASRI), and the In-Situ Groundwater Consulting (http://www.insituconsulting.co.za). Geological, topographical and land use GIS (Geographic Information Systems) data were obtained from the Water Resources of South Africa 2005 study (Middleton and Bailey, 2009).

The average catchment precipitation was obtained by studying seven weather stations with daily rainfall data from 2001 to 2012. Only four stations were selected based on data availability and proximity to the catchment. These stations were X1E006, X1E007, Barberton and Malelane (Figure 6-1). Missing rainfall values for Barberton (2%) and X1E007 (33%) were estimated by regression analysis. Malelane and X1E006 did not contain missing data. Using a Thiessen polygon distribution, the average rainfall was calculated for the catchment.

Average actual evaporation was calculated from daily Class A pan evaporation values from the Barberton and Malelane stations and daily Class S pan evaporation from X1E006 and X1E007 stations from 2003 to 2012. Class S pan evaporation was converted to Class A pan evaporation following the Water Resources of South Africa 1990 study WR90 (Midgley *et al.*, 1994). Class A evaporation was converted to reference evaporation using the guidelines for crop water requirements (Allen *et al.*, 1998) and reference evaporation was corrected for the specific land uses using data from the land satellite imagery collected from the Incomati Water Availability Assessment Study (DWAF, 2009b). Using a long term water balance from 2003 to 2012, actual mean evaporation rates were found.

To analyse the stream flow response at the outlet and tributaries, daily discharges at X2H022 (Outlet), X2H008 (Queens), X2H031 and X2H024 (Suidkaap) and X2H010 (Noordkaap) stream gauges were obtained from the DWS. The locations of the stations are shown in Figure 6-2A.

6.3.2 Field and laboratory methods

6.3.2.1 General

A field campaign from 20 November 2013 to 4 February 2014 was carried out to obtain an overview of the hydrochemistry of the catchment prior to the rainy season and to collect data for hydrograph separation studies.

Stream discharge collected from DWS data loggers (water levels converted to stream discharge using DWS rating curve) were retrieved at the outlet with a frequency of 12 minutes (0.2 hours) from 30 October 2013 to 17 February 2014. Hourly precipitation rates were obtained from the Incomati Catchment Management Agency (ICMA) rain gauges at Koffiekultuur, Nelshoogte Bos, Satico, and Josefdal Boarder from 1 October 2013 to 28 February 2014 (see locations on Figure 6-2A).

6.3.2.2 Water samples

Water samples were collected from the tributaries, the main river, one spring, and two drinking water wells as shown in Figure 6-2A. Each location was sampled twice during dry weather conditions. Each sample of approximately 250 mL was collected in polyethylene bottles, rinsed three times before the final sample was taken to avoid contamination, and refrigerated for sample preservation. Electrical conductivity (EC), pH and temperature were measured in-situ using a Wissenschaftlich-Technische-Werkstätten (WTW) conductivity meter.

6.3.2.3 Rain sampling

To obtain the isotopic and hydrochemical reference of rainfall, bulk rain samples were collected in the upstream and downstream part of the catchment. The rain samplers were constructed according to standards of the to avoid re-evaporation (Gröning *et al.*, 2012). Thus, an average of upstream and downstream samples per rain event was used for the rainfall end-member concentrations for each hydrograph separation.

Rainfall characteristics, including duration, total rainfall amount, maximum and average intensity, and the Antecedent Precipitation Index (API) were estimated for each rain event. A rainfall event was defined as a rainfall occurrence with rainfall intensity greater than 1 mm h^{-1}, and intermittence less than 4 h, as observed in a similar study in a semi-arid area by Wenninger *et al.* (2008). The API for n days prior the event were calculated using Equation (6.1):

$$API_{-n} = \sum_{i=1}^{7} P_{(-n-1+i)}(0.1i) \qquad [6.1]$$

where P in (mm h^{-1}) stands for precipitation and i is the number corresponding to the day of rainfall. For this study, APIs were calculated for the 7, 14, and 30 days prior to the event. Peak flow, runoff depth, and time to peak were determined for each event.

6.3.2.4 Automatic sampler

During the rainy season 2013-2014, four events that occurred on 12-13 December 2013 (Event 1), 28-30 December 2013 (Event 2), 13 January 2014 (Event 3) and 30-31 January 2014 (Event 4) were sampled using an automatic sampler manufactured by the University of KwaZulu-Natal (UKZN). The first two events were sampled on a volumetric basis obtaining 22 samples for Event 1, and 5 samples for Event 2 (a smaller number of samples were obtained for Event 2 due to photo sensor failure in the automatic sampler). Events 3 and 4 were sampled using a time-based strategy obtaining 13 samples for Event 3, and 36 samples for Event 4. A total volume of approximately 100 ml was obtained for each sample.

6.3.2.5 Chemical analysis of water samples

All samples were refrigerated, filtered, and analysed for HCO_3 and Cl using a Hach Digital Titrator, and SiO_2 using a Hach DR890 Portable colorimeter, within 48 h. Then, samples were transported to the IHE Delft laboratory in the Netherlands for further chemical analysis. The samples were analysed for major anions, cations and stable isotopes as listed in Table 6-1.

Table 6-1. IHE Delft laboratory equipment used in chemical analysis of Kaap catchment samples

	Parameter(s) analyzed	Equipment	Number of samples	Preservation method	Analytical uncertainty (σ)
Environmental Isotopes	^{18}O, ^{2}H	LRG DLT- 100 Isotope Analyzer	116	None	±0.2, ±1.5 (‰)
Cations	Ca^{2+}, Mg^{2+}, Na^{+}, K^{+}	Thermo Fisher Scientific XSeries 2 ICP -MS	116	Nitric acid (HNO_3)	± 0.2 (mg L^{-1})
Anions	Cl^{-}, NO_3^{-}-N, SO_4^{2-}, PO_4^{3-}	Dionex ICS-1000	116	Refrigerated at < 4 °C	± 0.2 (mg L^{-1})

6.3.3 Data analysis

6.3.3.1 Groundwater analysis

Groundwater chemical data for 240 boreholes and 18 borehole logs were obtained from In-Situ Groundwater Consultants covering the different geological formations (granite, lava, arenite, and gneiss). For 27 out of the 240 boreholes, pH, $CaCO_3$, Mg, Ca, Na, K, Cl, NO_3-N, F, SO_4, SiO_2, Al, Fe, and Mn data were available. The remaining boreholes only had information on EC, static water table depth, and physical characteristics of the borehole.

Borehole chemical data was classified according to the geological formations. The classified data distribution was observed using GIS, and basic statistical analysis was carried out to determine the control of geology over the hydrochemistry of groundwater.

To gain better insights with regard to groundwater flow, piezometric lines were created using an Inversed Distance Weighted (IDW) interpolation of the static water tables from the boreholes.

6.3.3.2 End member mixing analysis (EMMA)

Suitable parameters for hydrograph separation were identified by creating mixing diagrams of EC (μS cm^{-1}), SiO_2, $CaCO_3$, Cl, SO_4, Na, Mg, K, Ca (in mg L^{-1}) and δ^2H and $\delta^{18}O$ (‰VSMOW). Parameters were plotted against discharge to observe dilution and hysteresis effects. A principal component analysis was carried out based on the method described by Christophersen and Hooper (1992). Only not statistical correlated parameters were used. From these, the possibility of three end members was explored. The three runoff components identified were direct runoff, deep groundwater and shallow groundwater. Direct runoff was defined according to the conceptual model by Uhlenbrook and Leibundgut (2000) where direct runoff (or quick runoff component) was generated from direct precipitation on the stream channel, and overland flow from sealed and saturated areas and from highly fractured outcrops. The deep groundwater component was considered to be the portion of runoff generated from deep highly weathered granite aquifers, and the shallow groundwater component was considered to be the intermediate component from perched groundwater tables.

The mixing plot for δ^2H and K is presented in Figure 6-10. The direct runoff end member was characterized by the upstream and downstream rain samples. Potassium was used as an indicator of the shallow groundwater component due to the main sources of potassium, which are the weathering of minerals from silicate rocks, application of fertilizers, and the decomposition of organic material. The mobilization of potassium is linked to the flushing of the soil and shallow subsurface layers of vegetated areas. That was also observed by Winston and Criss (2002). The direct runoff samples had a low K average (0.5 mg L^{-1}) and depleted $\delta^{18}O$ and δ^2H values (-4.8‰ for $\delta^{18}O$; -27.5‰ for δ^2H). A spring sample was used as a proxy to characterize the deep groundwater component which contained more enriched $\delta^{18}O$ and δ^2H values (-0.9‰ for $\delta^{18}O$; -2.2‰ for δ^2H) and low K concentration (0.7 mg L^{-1}). The shallow groundwater end member was estimated considering the high K concentrations (4 mg L^{-1}) and slightly less depleted $\delta^{18}O$ and δ^2H (-3.5‰ for $\delta^{18}O$; -7.0‰ for δ^2H) observed in the stream samples. The error interval for the direct runoff in Figure 6-10 is ± the standard deviation of the rain samples. For the groundwater end-members, the error intervals were estimated as ±10% of the measured values.

While these errors are arbitrary, they were chosen as they are more conservative than the alternative analytical errors of ±0.2 mg L^{-1} for K and 1.5‰ for δ^2H and because there were no additional samples from which to derive the standard deviation.

6.3.3.3 Hydrograph separation

Isotope and hydrochemical data were combined with discharge data to perform a multi-component hydrograph separation based on steady state mass balance equations as described, for instance, in Uhlenbrook *et al.* (2002). The number of tracers (n-1) was dependent on the number of runoff components (n). Equations (6.2) and (6.3) were applied in dividing the total runoff, Q_T, into two and three runoff components.

$$Q_T = Q_1 + Q_2 .... + Q_n \quad [6.2]$$

$$c_T\, Q_T = c_1\, Q_1 + c_2 Q_2 ... + c_n Q_n \quad [6.3]$$

where Q_1, Q_2 and Q_n are the runoff components in $m^3\ s^{-1}$ and c_T, c_1, c_2 and c_n are the concentrations of total runoff, and runoff components.

6.3.3.4 Uncertainty estimation

Uhlenbrook and Hoeg (2003) showed that during the quantification of runoff components, uncertainties due to tracer and analytical measurements, intra-storm variability, elevation and temperature, solution of minerals, and the spatial heterogeneity of the parameter concentrations occur. For the Kaap River hydrograph separations, these uncertainties were accounted by the spatial hydrochemical characterization of the catchment and by sampling rainfall during each event and at different locations. Moreover, tracer end-members and analytical uncertainties were estimated using a Gaussian error propagation technique and a confidence interval of 70% as described by Genereux (1998) and Liu *et al.* (2004).

$$W = \left\{ \left[\frac{\partial y}{\partial x_1} W_{x1} \right]^2 + \left[\frac{\partial y}{\partial x_2} W_{x2} \right]^2 + \square + \left[\frac{\partial y}{\partial x_n} W_{xs} \right]^2 \right\}^{\frac{1}{2}} \quad [6.4]$$

W is the estimated uncertainty of each runoff component (e.g. direct runoff, shallow and deep groundwater components). W_{x1} and W_{x2} are the standard deviations of the end-members. W_{xs} is the analytical uncertainty and the partial derivatives $\frac{\partial y}{\partial x_1}$, $\frac{\partial y}{\partial x_2}$ and $\frac{\partial y}{\partial x_n}$ are the uncertainties of the runoff component contributions with respect to the tracer concentrations.

6.4 Results

6.4.1 Hydrology, hydrogeochemistry and groundwater flow

One of the characteristics of semi-arid areas is the high variability of flows. This large variability is observed at the Kaap River outlet and tributaries (Table 6-2), where the highest and lowest flows recorded at the Kaap outlet are 483 and 0 m^3 s^{-1}, respectively. Pardé coefficients (Figure 6-3) reflect the seasonal flow behavior showing the dominance of one rainy season per hydrological year with the largest flows occurring in February. Moreover, the flat slopes observed at the upper end of the flow duration curves (Figure 6-4) are evidence of groundwater storage areas located in the upstream part of the catchment.

Table 6-2. Physical and hydrological characteristics of the Kaap tributaries and outlet

Tributary name	**Kaap outlet**	**Queens**	**Upper Suidkaap**	**Noordkaap**
Station ID	X2H022	X2H008	X2H031	X2H010
Reach length (km)	45.7	41.2	42.5	57.5
Sub-basin area (km^2)	1640	291	256	315
Data analyzed	1961–2012	1949–2012	1967–2012	1970–2012
Period (years)	51	63	45	42
% data missing	5%	0%	3%	6%
Highest flow measured (m^3 s^{-1})	483	96	123	28
Lowest flow measured (m^3 s^{-1})	0	0	0	0
Mean of yearly highest flows (m^3 s^{-1})	65	13	19	6
Mean of yearly lowest flows (m^3 s^{-1})	0.4	0.1	0.3	0.2
Mean flow MQ (m^3 s^{-1})	3.6	0.6	1.1	0.6
Variability ratio	180	186	65	31
Specific discharge (L s^{-1} Km^{-1})	3	2.2	4.2	1.9
Maximum and average days of no flow per year	139; 8	12; 1	23; 1	17; 1

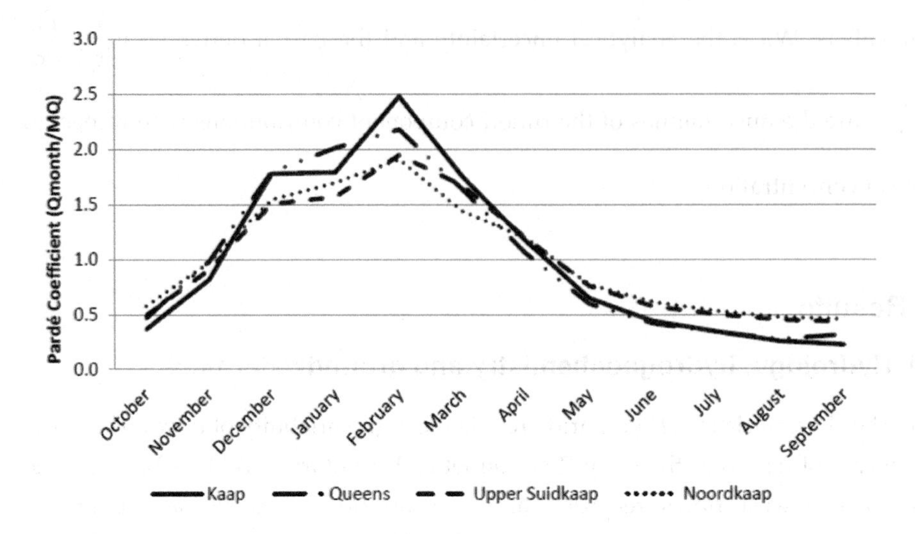

Figure 6-3. Annual flow regimes at X2H022 (outlet), X2H008 (Queens), X2H031 and X2H024 (Suidkaap) and X2H010 (Noordkaap) based on long-term flow data.

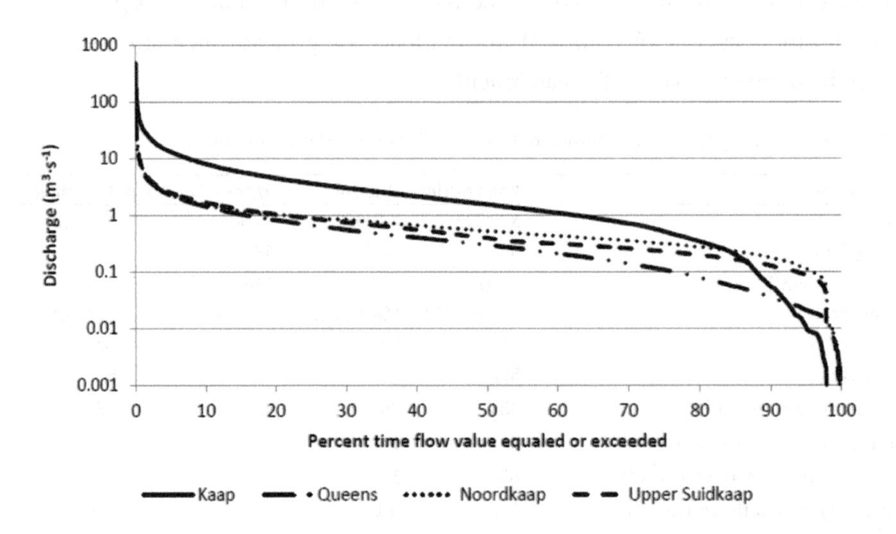

Figure 6-4. Flow duration curves for X2H022 (outlet), X2H008 (Queens), X2H031 and X2H024 (Suidkaap) and X2H010 (Noordkaap) based on long-term flow data.

The variability of the catchment's groundwater quality parameters was studied from borehole data. Electrical conductivities in the granite region had the lowest electrical conductivity (EC) values (average 383 μS cm^{-1}), while the gneiss formation, near the outlet, had the largest EC average of 1140 μS cm^{-1}. Lava and arenite formations had mean EC values of 938 and 525 μS cm^{-1}, respectively. The gneiss and lava formations had higher concentration averages of chloride and calcium carbonate than the granite and arenite formations. These can be seen in the box plots in Figure 6-5.

Groundwater piezometric lines followed the topographical relief. The highest water tables were observed at the northern boundary of the catchment, with water tables up to 1150 m (Figure 6-2D). From the groundwater piezometric map, it was observed that groundwater moves towards the stream, indicating a gaining river system. Time series data from boreholes did not show a significant change in water tables due to seasonal or long-term changes.

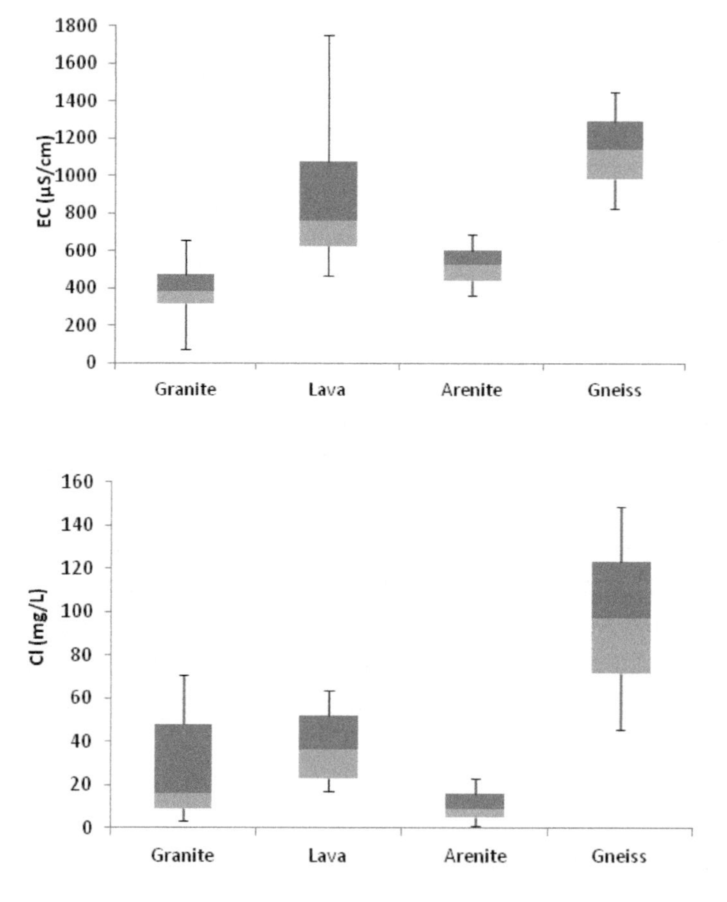

Figure 6-5. Boxplots of borehole water quality parameters at different geological locations in the Kaap catchment.

6.4.2 Spatial hydrochemical characterization

The upstream rain sample average had a more depleted isotopic signature (-5.1‰ for $\delta^{18}O$; -30.2‰ for $\delta^{2}H$) than the lower-elevation rain sample average (-4.4‰ for $\delta^{18}O$; -4.7‰ for $\delta^{2}H$). Upstream and downstream delta deuterium values ranged from a minimum of -30.2‰to a maximum of -21.8‰ and delta oxygen-18 ranged from -5.14 to -3.72‰. Baseflow at the catchment outlet (X2H022) was characterized by analyzing DWS long-term water quality data and by field sampling prior to the 2013-2014 rainy season. Table 6-3 shows the results from the field sampling.

Table 6-3. List of mean values of hydrochemical parameters obtained during the 2013-2014 field campaign

	Location			
Parameter	**Suidkaap**	**Queens**	**Noordkaap**	**Outlet**
EC ($\mu S\ cm^{-1}$)	84.0	128.7	92.9	443.0
SiO_2 ($mg\ l^{-1}$)	22.4	17.0	20.9	24.1
$CaCO_3$ ($mg\ l^{-1}$)	38.5	59.5	41.3	154.0
Cl ($mg\ l^{-1}$)	3.8	3.6	2.8	15.5
SO_4 ($mg\ l^{-1}$)	1.8	4.1	1.6	47.2
Na ($mg\ l^{-1}$)	7.5	7.1	7.3	29.3
Mg ($mg\ l^{-1}$)	2.8	7.4	3.7	25.3
Ca ($mg\ l^{-1}$)	7.9	9.1	6.8	27.6
$\delta^{2}H$ (‰ VSMOW)	-12.1	-12.4	-12.7	-8.9
$\delta^{18}O$ (‰ VSMOW)	-3.2	-3.1	-3.5	-2.7

The upper section of the catchment, mainly dominated by granite, is characterized by low to moderate electrical conductivities. Long-term mean electrical conductivities (sampled monthly by the DWS from 1984 to 2012) for the Upper Suidkaap and Noordkaap tributaries were 75 and 104 $\mu S\ cm^{-1}$, respectively. On the contrary, the catchment outlet had a higher long-term average EC of 572 $\mu S\ cm^{-1}$ (DWS long-term monthly average from 1977 to 2012).

6.4.3 Rainfall-runoff observations

Table 6-4 summarizes the rainfall-runoff observations for the four studied events. The events had distinctive characteristics showing large variability in peak flows, API, rainfall duration, rain depth and maximum and average intensities. Event 1 had the highest peak flow at 124 $m^3\ s^{-1}$ while Event 3 had the smallest peak flow at 6.5 m^3

s^{-1}. APIs, especially API-7, differed from very wet conditions during Event 1 (39 mm) to very dry conditions (1 mm) during Event 2. Event 1 was a relatively short event (7 h) with high antecedent precipitation conditions and high rainfall intensities generating the largest amount of runoff at the outlet. In contrast, Event 3 was a short event with average rainfall intensity that generated the lowest peak flow.

Table 6-4. Rainfall-runoff relationships observed during the 2013-2014 wet season for Kaap catchment at outlet (X2H022 stream gauge) and average precipitation from Koffiekultuur, Nelshoogte Bos, Satico, and Josefdal Boarder rain stations

		Event 1	Event 2	Event 3	Event 4
Runoff	Peak flow time and date	13 Dec 2013, 18:24	30 Dec 2013, 6:12	16 Jan 2014, 3:48	31 Jan 2014, 17:00
	Maximum river depth (m)	2.0	1.0	0.5	0.5
	Peak flow (m^3 s^{-1})	124.0	27.6	6.5	7.1
	Runoff Volume (mm)	3.2	2.6	0.1	0.4
	Time to peak after rainfall started (hrs)	24.4	31.2	60.8	22.0
Rainfall	Rainfall start date & time	12 Dec 2013, 18:00	28 Dec 2013, 23:00	13 Jan 2014, 15:00	30 Jan 2014, 19:00
	Rainfall duration (hrs)	7	39	7	26
	Rainfall depth (mm)	24	78	17	20
	Average rainfall intensity (mm hr^{-1})	3.4	2.0	2.5	0.8
	Maximum rainfall intensity (mm hr^{-1})	9.8	12	5	10
	Antecedent Precipitation Index API_{-7} (mm)	38.7	1.3	7.8	24.9
	API_{-14} (mm)	118.1	12.8	20.0	67.9
	API_{-30} (mm)	390.2	220.8	192.4	223.8

6.4.4 Response of isotopes and hydrochemical parameters

During the storm events, most hydrochemical parameters (EC, Ca, Mg, Na, SiO_2 and Cl) and water isotopes (δ^2H and $\delta^{18}O$) showed dilution responses except for potassium (Figure 6-6). The first flood was the largest event sampled, reaching a peak flow of 124 m^3 s^{-1} where a large contribution of direct runoff was observed. In this event, a large degree of dilution of the sampled hydrochemical parameters is also observed. The following events had smaller peak flows of 27.6, 6.5 and 7.1 m^3 s^{-1} for events 2, 3 and 4 respectively. Thus, smaller dilution effects were observed for events 2, 3 and 4. The smaller peak flows and lower direct runoff contributions for the latter events may explain the temporal variability observed in the increased concentrations of the hydrochemical parameters over time. During Event 1, EC's initial value of 317 μS cm^{-1} decreased to 247 μS cm^{-1} during peak flow. Similarly, $CaCO_3$ and SiO_2 decreased from 115 to 82 mg L^{-1} and from 21.1 to 19.6 mg L^{-1}, respectively. $\delta^{18}O$ (-2.9‰) and $\delta^{2}H$ (-7.0‰) decreased to -3.2 and -12.6‰, respectively. Potassium concentrations increased from 1.3 to 2.8 mg L^{-1}. For Event 2, a smaller number of samples were collected due to malfunctioning of the automatic sampler. However, dilution of SiO_2 and Cl, and an increase in potassium concentrations were observed. Event 3 and 4 were relatively small events, but showed the same dilution behaviour of the sampled parameters and the increase in potassium concentrations.

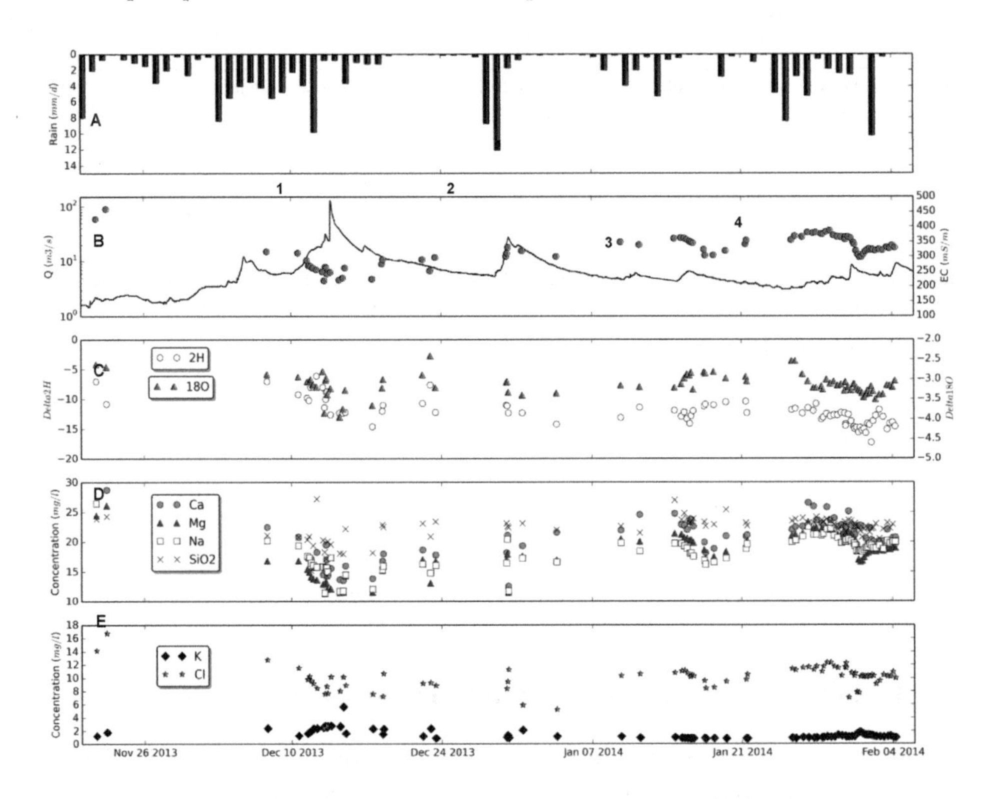

Figure 6-6. Kaap catchment: a) average precipitation in mm day^{-1}, b) discharge at the outlet in m^3 s^{-1} and electrical conductivity μS cm^{-1}, c) delta deuterium and delta oxygen-18 in ‰ VSMOW, d) calcium, magnesium, sodium and silica concentrations at the outlet in mg L^{-1}, and e) chloride and potassium concentrations at the outlet in mg L^{-1}.

6.4.5 Two-component hydrograph separation

Event and pre-event components were separated using $\delta^{18}O$ and $\delta^{2}H$, and direct runoff and groundwater were separated using EC, SiO_2, $CaCO_3$, and Mg. For simplicity, the two hydrograph separation components in this study are referred to as direct runoff and groundwater components. Direct runoff (quick flow component) defined in the methods section as the portion of direct precipitation and infiltration excess overland flow, was characterized using the rainfall samples collected upstream and downstream inside the catchment. Groundwater end-members were obtained from the initial stream water samples before the rainfall started. Events 1 and 4 had the largest contributions of direct runoff among the four events accounting for 29% in case of Event 1 and up to 36% for Event 4 (Table 6-5). Events 2 and 3 had lower direct runoff contributions ranging from 5 to 13% for Event 2 and 2 to 12% for Event 3. Figure 6-7 presents the two-component hydrograph separations for the four events.

Table 6-5. Percentages of direct runoff [DR] and groundwater [GW] contributions and 70% uncertainty percentages [W] from two-component hydrograph separations for the 2013-2014 wet season Kaap catchment, South Africa.

Tracer	Event 1			Event 2			Event 3			Event 4		
	DR	GW	W	DR	GW	W	DR	GW	W	DR	GW	W
EC	22	78	6.8	5	95	7.9	6	94	7.0	27	73	4.2
SiO_2	21	79	2.6	6	94	2.5	12	88	2.2	21	79	2.6
$CaCO_3$	29	71	6.3	9	91	6.9	6	94	6.8	24	76	4.6
Mg	22	78	5.6	13	87	6.0	8	92	5.3	24	76	4.0
^{18}O	23	77	8.6	8	92	3.3	10	90	3.1	36	64	12.4
^{2}H	19	81	5.6	5	95	15.0	2	98	19.4	21	79	24.9

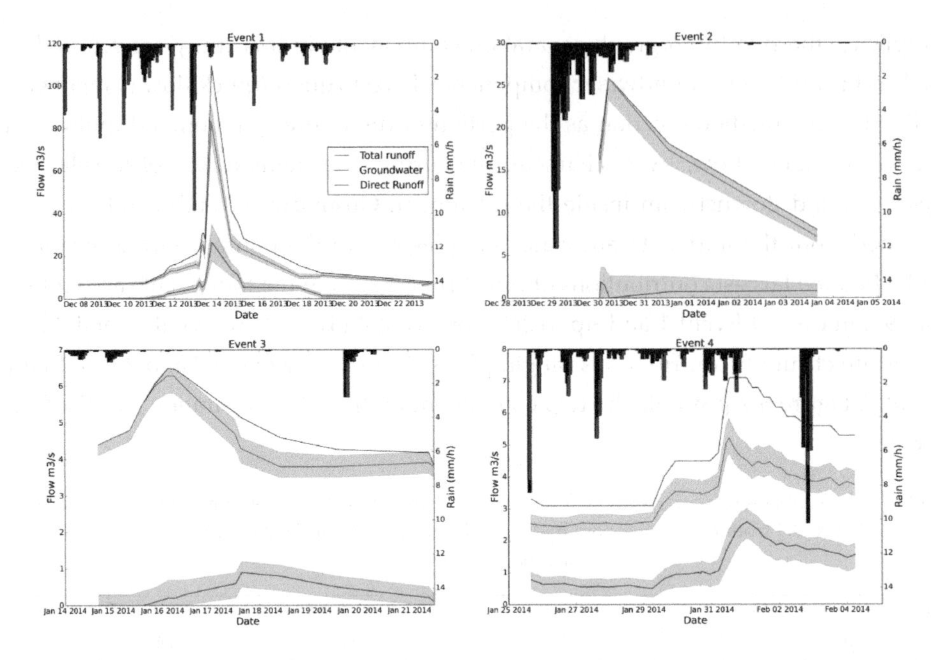

Figure 6-7. Two component hydrograph separations using electrical conductivity as a tracer. Event 1 and 4 had larger direct runoff contribution coinciding with the total runoff peak. Event 2 and 3 had smaller direct runoff contribution

6.4.6 Isotope hydrograph separation versus hydrochemical hydrograph separation

Hydrochemical tracers usually separate runoff from source areas, while isotopes generally separate old water from new water. The definition of Klaus and McDonnell (2013) was used for this study, stating that pre-event water (or old water as referred to in this section of the study) is the water stored in the catchment before the rainfall event. This component may not be representative of deep groundwater sources, but it may be water stored from the same rainfall season but from previous rainfall events. Thus, a comparison between "oldwater" and "groundwater" components obtained during the four events was carried out to investigate to what extent these components are similar. This allowed us to determine the suitability of isotopic hydrograph separations versus hydrochemical separations for semi-arid environments. Figure 6-8 presents the percentages of groundwater and old water contributions using environmental isotopes (δ^2H and $\delta^{18}O$) and hydrochemical (EC, SiO_2, $CaCO_3$ and Mg) tracers for the four investigated events. It is noted that Events 1 and 4 have smaller contributions of groundwater than Events 2 and 3. During Event 4 and Event 2, old water resembles groundwater. The data points above the line present instances where old water is not necessarily groundwater, but water stored before the event. No major differences are observed from using hydrochemical or isotope tracers for the hydrograph separation.

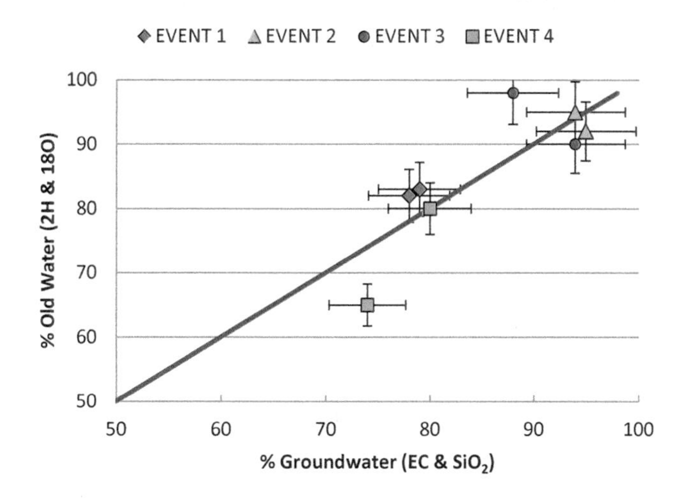

Figure 6-8. Percentages of groundwater and old water contributions using environmental isotopes (δ^2H and $\delta^{18}O$) and hydrochemical (EC and SiO_2) tracers

6.4.7 End-member mixing analysis (EMMA)

To further differentiate the runoff components, a Principal Component Analysis (PCA) was carried out on 12 solutes (EC, SiO_2, $CaCO_3$, Cl, NO_3-N, SO_4, Na, Mg, K, Ca, $\delta^{18}O$ and δ^2H) using the R statistical software (R Development Core Team, 2014). The correlation matrix was used for the PCA. Results indicated that 90% of the variability is explained by two principal components (m). Thus, the number of end-members (n) can be chosen as ($n=m+1$) leading to a three component hydrograph separation (Christophersen and Hooper, 1992). Figure 6-9 shows the biplot of principal components where the orthogonal vectors indicate no dependency between parameters. This is observed for $\delta^{18}O$, δ^2H, K, and NO_3. The clustering of the hydrochemical parameters reveals the strong correlation between these parameters (SiO_2, $CaCO_3$, Ca, EC, Mg, Na, Cl, and SO_4). Potassium shows a negative strong correlation with the clustered parameters but not with the water isotopes and NO_3. Thus, for the three component hydrograph separations, orthogonal vectors with weak Pearson correlations were selected. These are K and $\delta^{18}O$ ($r = -0.28$) and K and δ^2H ($r = 0.45$). The latter shown in Figure 6-10. Nitrate was not selected due to its non-conservative properties. Potassium was identified as a useful tracer due to its increasing concentrations during runoff peaks. This high potassium concentration suggested the presence of soil water influenced by mobilization of fertilizer and/or organic material. To account for additional near-surface water, this component is referred to as the "shallow groundwater component" during this study. It is important to note that the shallow groundwater component could be a mix of surface runoff and near-surface water since potassium was used as an indicator of shallow groundwater, and this element can also be found in surface runoff.

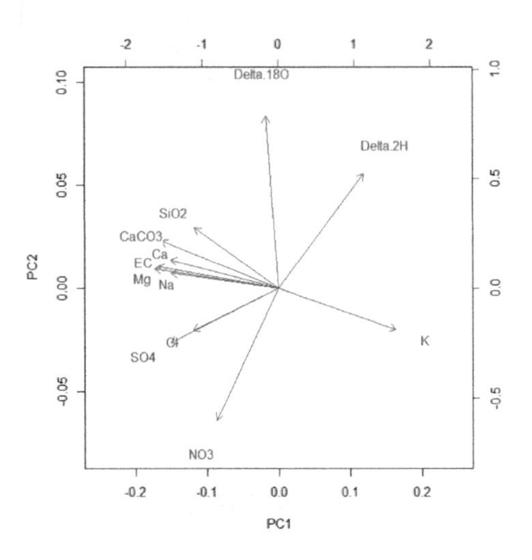

Figure 6-9. Biplot of principal components generated during PCA analysis of stream water samples using EC, SiO_2, $CaCO_3$, Cl, NO_3-N, SO_4, Na, Mg, K, Ca, δ^2H, $\delta^{18}O$.

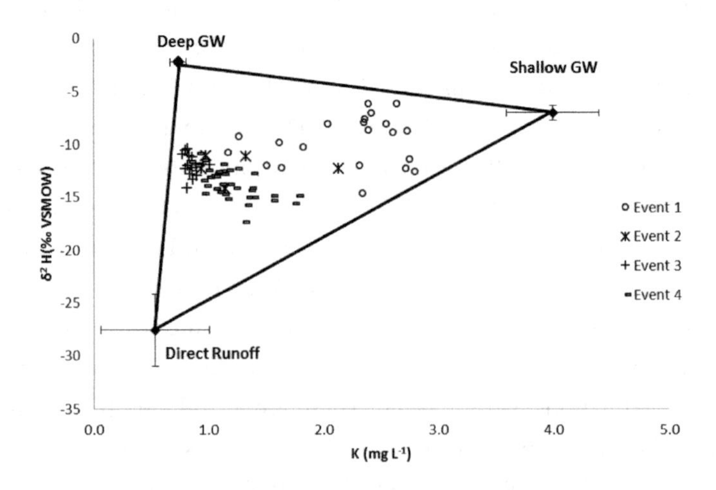

Figure 6-10. Mixing diagram of δ^2H and K showing stream water samples at outlet for four rain events during the 2013-2014 wet season.

6.4.8 Three-component hydrograph separation

Direct runoff contributions obtained during the three-component hydrograph separations (Table 6-6 and Figure 6-11) concur with the two-component hydrograph separations. Events 1 and 4 were characterized by higher contributions of direct runoff than events 2 and 3. Moreover, Event 1 also had a higher contribution of shallow groundwater that peaked during the total runoff peak. Events 2, 3 and 4 had higher deep groundwater contributions. Uncertainties for the three-component hydrograph separations can be seen in Table 6-6.

Table 6-6. Direct runoff [DR], shallow groundwater [SGW], and deep groundwater [DGW] contributions in (%) and 70% uncertainty of 3-component hydrograph separations in (%)

Tracers	Event 1			Event 2			Event 3			Event 4		
	DR	SGW	DGW	DR	SGW	DGW	DR	SGW	DGW	DR	SGW	DGW
K &^{18}O	28	45	26	7	19	74	16	6	78	41	21	37
70% uncertainty (%)	7.2	5.3	5.8	7.4	3.2	5.1	5.3	3.0	3.9	7.9	6.2	5.8
K &^{2}H	22	45	33	14	19	67	11	5	84	37	20	42
70% uncertainty (%)	4.8	6.6	6.4	3.8	3.9	5.5	3.0	2.8	4.0	6.3	6.2	7.6

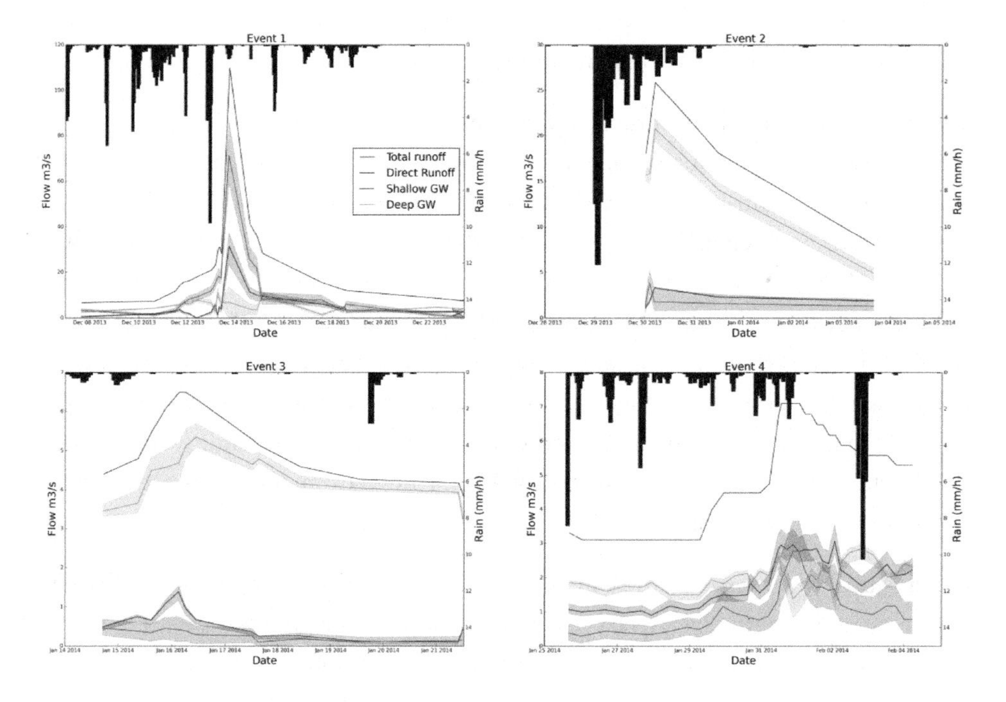

Figure 6-11. Three-component hydrograph separations using K and ^{2}H.

6.5 Discussion

6.5.1 Runoff processes in the Kaap catchment

From the mixing diagrams, groundwater analysis and spatial hydrochemical characterization of the catchment, the runoff components were identified and characterized. The groundwater analysis suggested two sources of groundwater of different ionic content at the upper and lower sections of the catchment. In the upstream area, granite is the dominant formation explaining the lower ionic content in groundwater in contrast to the downstream areas where geologically diverse formations and land use increase the ionic content of groundwater. The weathered granite layer allows rain to infiltrate to the deeper groundwater reservoir through preferential flow paths with less contact time for weathering processes to occur. This explains the hydrochemical signature of the deep groundwater component, which is characterized by its moderate electrical conductivities, moderate to high dissolved silica, lower ionic content and low potassium concentrations. The chemical signature of the shallow groundwater component is characterized by the high electrical conductivities, alkalinity, sulfates, potassium, and nitrates which are washed from top geological layers with large ionic content and land uses such as agriculture and mining which are more predominant in the downstream region of the catchment.

The three-component hydrograph separations suggest that the shallow groundwater component (potentially including surface runoff) is quickly activated during rainfall events, and its contribution increases as the antecedent precipitation increases as observed during Events 1 and 4, where the shallow groundwater contributions were 45 and 20–21 %, respectively. Moreover, a connection between surface and groundwater is evident from the groundwater piezometric map (Figure 6-2D), which shows a gaining river system, and from the flow duration curves, which indicate exfiltrating groundwater storages to the streams. Further literature (Hughes, 2010) suggests that most of South Africa's groundwater is stored in secondary aquifers and that surface flow may be nourished by lateral flow from semi-saturated fracture systems after storm events.

Other studies (Petersen, 2012) in the nearby Kruger National Park (KNP) have shown that groundwater recharge occurs mostly during the wet season and groundwater flow travels in accordance with the topographical relief. Petersen (2012) studied a granite-dominated area and a basaltic rock-dominated area, approximately 30 km east from the Kaap outlet. The study found that the granite region was mainly characterized by the steep topography which favors overland flow which infiltrates through depressions, cracks and fractures by preferential pathways, while the southern basaltic section with a flatter topography showed piston flow processes to be more predominant. The Petersen (2012) findings, covering studies of approximately 1011 boreholes in KNP, support the findings in the Kaap catchment where high fracturing in the granite section allows recharge of deeper groundwater reservoirs through preferential flowpaths.

It is important to note that the inferences drawn from this study are based on four events sampled during the 2013-2014 wet season but supported by historical meteorological, hydrological and water quality data, groundwater analysis and a spatial hydrochemical study of the catchment. In addition, Table 6-7 shows runoff studies with similar number of events studied.

Table 6-7. Runoff studies with number of events studied

Study name	**Reference**	**Number of events**
Hydrograph separation using stable isotopes, silica and electrical conductivity: an alpine example	Laudon and Slaymaker (1997)	5
The role of soil water in stormflow generation in a forested headwater catchment: synthesis of natural tracer and hydrometric evidence	Bazemore *et al.* (1994)	2
Quantifying contributions to storm runoff through end-member mixing analysis and hydrologic measurements at the Panola Mountain Research Watershed (Georgia, USA)	Burns *et al.* (2001)	2
On the value of combined event runoff and tracer analysis to improve understanding of catchment functioning in a data-scarce semi-arid area	Hrachowitz *et al.* (2011)	28
Quantifying uncertainties in tracer-based hydrograph separations: a case study for two-, three- and five-component hydrograph separations in a mountainous catchment	Uhlenbrook and Hoeg (2003)	4
Hydrograph separations in a mesoscale mountainous basin at event and seasonal timescales	Uhlenbrook *et al.* (2002)	2
Identification of runoff generation processes using combined hydrometric, tracer and geophysical methods in a headwater catchment in South Africa	Wenninger *et al.* (2008)	3
Runoff generation in a steep, tropical montane cloud forest catchment on permeable volcanic substrate	Muñoz-Villers and McDonnell (2012)	13
Quantifying the relative contributions of riparian and hillslope zones to catchment runoff	McGlynn and McDonnell (2003)	2
Dynamics of nitrate and chloride during storm events in agricultural catchments with different subsurface drainage intensity (Indiana,	Kennedy *et al.* (2012)	2

USA)		
Investigation of hydrological processes using chemical and isotopic tracers in a mesoscale Mediterranean forested catchment during autumn recharge	Marc *et al.* (2001)	3

6.5.2 The catchment's response dependency on antecedent precipitation

Hydrograph separation results suggested that there is a direct runoff contribution (2-36%) to total runoff during storm events for the Kaap River. Similar results have been obtained for other catchments in semi-arid areas. For instance, Hrachowitz *et al.* (2011) in their study in four nested catchments in Tanzania found event runoff coefficients of 0.09. Similarly, Munyaneza *et al.* (2012) found groundwater contributions up to 80% of total runoff in the Mingina catchment in Rwanda using the two and three-hydrograph separations in a 258 km^2 catchment. The importance of sub-surface flow in semi-arid catchments is also illustrated in Wenninger *et al.* (2008) in the Weatherley catchment in the Eastern Cape in South Africa.

From the several variables considered such as geology, topography and rainfall characteristics studied for the four events, the direct runoff component was most sensitive to the API. This is observed during Events 1 and 4 where API_{-7} values are the largest among the four events and direct runoff contributions are also the largest for these events. The relationship between API_{-7} and direct runoff generation is supported by a strong Pearson correlation (0.76-0.94). This suggests that direct runoff is enhanced by wetter conditions in the catchment due to saturation in the subsurface triggering saturation overland flow.

6.5.3 Complexities of runoff processes understanding in semi-arid areas

The combination of climatic and hydrological processes influenced by topography, geology, soils and land use make catchments complex systems. Although the opposite may be true for particular situations, in general, catchments become more non-linear as aridity increases and runoff processes become more spatially and temporally heterogeneous than in humid regions (Farmer *et al.*, 2003; Blöschl *et al.*, 2013a). Thus, understanding hydrological processes in arid catchments becomes more difficult due to high variability of rainfall and streamflow, high evaporation losses, long infiltration pathways, permeable stream channel beds and often deep groundwater reservoirs (Hughes, 2007; Trambauer *et al.*, 2013).

The high variability of rainfall increases difficulties of runoff prediction by triggering different runoff responses. For instance, high intensity storms tend to generate overland flow in the form of infiltration excess overland flow (Smith and Goodrich, 2005), while high antecedent precipitation conditions enhance saturation excess overland flow. This effect is visible in this study during Event 1 where the high API suggested saturation of the subsurface, thus reducing the infiltration capacity and enhancing saturation excess overland flow. The opposite is observed for Events 2 and 3, where the low soil moisture conditions allow more rainfall to infiltrate activating other runoff processes such as preferential vertical flow.

Although not included in this study, inter-annual variability, evaporation, hydraulic connectivity, permeable stream beds and interception have shown to change the behavior of runoff processes in arid and semi-arid areas. For instance, inter-annual rainfall variability is closely related to high evaporation losses. Mostert *et al.* (1993) study in a Namibian Basin found that during wetter seasons, vegetation cover and total evaporation increased, thus reducing the amount of runoff reaching the outlet. Similarly, hydraulic connectivity in arid environments is limited by the reduced soil moisture conditions in these areas, leading to reduced groundwater recharge. Other fluxes such as interception and flow through permeable stream beds pose a greater challenge to the understanding of runoff processes in semi-arid areas. Interception can further decrease the hydrologic connectivity, breaking the link between meteoric water and groundwater as observed in the Zhulube catchment in Zimbabwe, where interception accounted up to 56% of rainfall during the dry season (Love *et al.*, 2010a). Similarly, transmission losses due to the high degree of fracturing of stream beds can significantly reduce streamflow but increase recharge of groundwater systems.

Thus, this study illustrated the effects of temporal rainfall variability during the wet season, suggesting the influence of antecedent precipitation conditions on direct runoff generation. However, studying the effects of spatial and inter-annual rainfall variability, high evaporation and transpiration (from unsaturated zones, alluvial aquifers, and riparian zones) fluxes, the spatial variability of vegetation, and deep groundwater resources on streamflow generation is still required for better understanding of runoff processes in semi-arid areas. More monitoring of groundwater levels and aquifers would assist in bridging this gap of knowledge, such as in Van Wyk *et al.* (2012). Emphasis is placed on studying the region during dry weather for further understanding of evaporation and transpiration from deeper layers of soil moisture that in some cases can reach even into groundwater systems (e.g., eucalyptus trees).

6.6 Conclusions

The Kaap catchment has suffered devastating floods that affect greatly the trans-boundary Incomati basin, in particular downstream areas in South Africa, Swaziland and Mozambique, where recent floods have caused significant economic and social losses. Runoff processes were poorly understood in the Kaap catchment, limiting rainfall-runoff models to lead to better informed water management decisions. Through hydrometric measurements, tracers and groundwater observations, runoff components and main runoff generation processes were identified and quantified in the Kaap catchment for the 2013-2014 wet season. The suitability of isotope hydrograph separation was tested by comparing it to hydrochemical hydrograph separations showing no major differences between these tracers. Hydrograph separations showed that groundwater was the dominant runoff component for the wet season 2013-2014. Three component hydrograph separations suggested a third component that we addressed as the shallow groundwater component. However, further research is still necessary to make a clear distinction between surface runoff and shallow groundwater. A strong correlation between direct runoff generation and antecedent precipitation conditions was found for the studied events. Direct runoff was enhanced by high antecedent precipitation activating saturation excess overland flow. Similar groundwater contributions have been observed in other studies in semi-arid areas (Wenninger *et al.*, 2008; Hrachowitz *et al.*, 2011; Munyaneza *et al.*, 2012). The understanding of runoff generation mechanisms in the Kaap catchment contributes to the limited number of hydrological processes studies and in particular hydrograph separation studies in semi-arid regions for the proper management of water resources. Moreover, this study was carried out during the wet season, and in order to gather a better understanding of the hydrological system, further studies focusing on the dry season are still needed, particularly on the dependency of runoff generation on soil moisture and vegetation.

7

HYDROLOGICAL MODELLING OF THE KAAP CATCHMENT

With the aim to better understand the key hydrological processes and runoff generation mechanisms in the semi-arid meso-scale Kaap catchment in South Africa, a hydrological model was developed using the open source STREAM model. Dominant runoff processes were mapped using a simplified Height Above the Nearest Drainage (HAND) approach combined with geology. Furthermore, the model was informed by process studies (hydrograph separation using tracers and digital filters) to enhance understanding on runoff processes in the Kaap. The Prediction in Ungauged Basins (PUB) framework of runoff signatures was used to analyse the model results. The results showed distinct patterns of flow generation in the Noordkaap and Suidkaap catchments, versus the Queens and Kaap. Furthermore, the high impact of water abstractions and evaporation during the dry season was highlighted, affecting low flows in the catchment. Results also indicate that the root zone storage and the parameters of effective rainfall separation (between unsaturated and saturated zone), quickflow coefficient and capillary rise, were very sensitive in the model. The inclusion of capillary rise (feedback from saturated to unsaturated zone) greatly improved simulation results, which suggests greater surface-groundwater interactions, particularly in the Queens and Kaap catchments.

This chapter is based on: Saraiva Okello AML, Masih I, Uhlenbrook S, Jewitt GPW, Van der Zaag P. 2018a. Improved Process Representation in the Simulation of the Hydrology of a Meso-Scale Semi-Arid Catchment. *Water*, 10: 1549. DOI: https://doi.org/10.3390/w10111549.

7.1 Introduction

In many regions of the world, including in most parts of Southern Africa, data availability and resources for detailed field investigations are limited (Hughes, 2016). Large catchments need to be modelled with limited input data, yet the results are needed to manage water resources that are crucial to livelihoods and the environment (Hughes *et al.*, 2015; Hughes, 2016). The importance of process understanding is critical (Blöschl *et al.*, 2013b; Hrachowitz *et al.*, 2013) because the results of modelling are used for every day water management - water resource availability assessments, water resource development, water releases from dams, environmental flow assessments, etc. (Hughes, 2016). Modellers therefore need to strike a balance between model complexity and data availability – even though more complex models strive to represent the various hydrological processes, they are more data intensive, and often such data are not readily available (Hughes, 2016). Therefore, approaches that rely on readily available data could provide better avenues to improve hydrological models and understanding of hydrological processes.

The Predictions in Ungauged Basins (PUB) decade of the International Association of Hydrological Sciences (IAHS) provided numerous tools and approaches to improve understanding of hydrological processes in data-limited environments. Blöschl *et al.* (2013b) present a useful summary of some of these approaches and suggest a framework to analyse hydrological processes through the use of runoff signatures (Blöschl *et al.*, 2013b; Viglione *et al.*, 2013). Runoff signatures are the temporal patterns of runoff response of catchments, derived from observed or modelled series of flow data (Viglione *et al.*, 2013). They are intended to extract relevant information about hydrological behaviour, such as to identify dominant processes, and to determine the strength, speed, and spatiotemporal variability of the rainfall–runoff response (McMillan *et al.*, 2017).

The literature suggests several approaches to define dominating runoff processes (DRP) zones. Some researchers make use of very detailed field investigations (e.g. Uhlenbrook, 2003), with intensive drilling, soil mapping, and interpretation of aerial images (Scherrer and Naef, 2003; van Tol *et al.*, 2013). Others make use of geological surveys, and hydrogeological maps. Some attempts have been made to use simplified approaches, only requiring readily available DEM, geological maps and land use maps (Müller *et al.*, 2009; Hümann and Müller, 2013). Recently, further simplification was suggested by Savenije (2010), by using only topography-derived information based on the co-evolution concept. This approach was tested in catchments in Luxembourg and Thailand (Gharari *et al.*, 2011; Gao *et al.*, 2014).

Savenije (2010) argues that catchments have "organized complexity", and therefore relatively simple models can perform well. Thus, it is necessary to model only dominant hydrological processes at the relevant scale using, for example, the concept of landscape zones e.g. plateaus, hillslopes and wetlands. These zones have different dominant hydrological processes, and can be defined by topographic indicators, such as Height Above the Nearest Drainage (HAND). According to Savenije (2010), plateaus are hydrological landscapes with modest slope and deep groundwater and where the dominant runoff mechanism is evaporation excess deep percolation. Plateaus are often used for agriculture and the vertical processes such as evaporation and recharge dominate. Wetlands and riparian zones are mostly dominated by saturation excess overland flow runoff mechanism. Hillslopes are hydrological landscapes where storage excess subsurface flow is the dominant mechanism. They are often covered by forest, and generate significant portions of runoff

In central and northern European landscapes, the hillslopes perform the functions of drainage and moisture retention (Savenije, 2010), in order to sustain the predominantly forest ecosystem. Subsurface drainage occurs through preferential pathways. Hillslopes also establish the subsurface connection to the groundwater storage of plateaus.

Soil can also be a first-order control in partitioning hydrological flow paths, residence times and distributions as well as water storage, particularly in smaller catchments (Soulsby *et al.*, 2006; van Zijl and Le Roux, 2014). In South Africa, the study of dominant runoff generation processes has been spearheaded by the hydropedology community (Van Huyssteen, 2008). Schulze (1985) was a pioneer in using pedo-transfer functions and decision support systems (Pike and Schulze, 1995) to derive relevant information for hydrological models from soil maps. van Tol *et al.* (2013) provide a useful framework for classification of hillslopes in South Africa, based on their soils/geology/hydropedology. It provides detailed explanations of how different hillslopes respond to rainfall, and how the water flows through them.

Several studies (e.g. Van Tol *et al.*, 2015; van Zijl *et al.*, 2016) have used digital soil mapping to derive parameters for hydrological models. However, detailed soils map information at the level of transboundary river basins, such as the Incomati basin is not available. The Soil Grids 250m database (Hengl *et al.*, 2015; Hengl *et al.*, 2017) was produced using digital soil mapping techniques, and could potentially fill this gap in making more detailed soil information readily available for hydrological modelling.

The majority of studies on hydrological processes and runoff generation processes are conducted in small-scale research catchments. However, the understanding of runoff processes is critical for informing models at meso/large catchment scales, where additional processes come into play (e.g. space-time variability of rainfall, runoff routing, larger heterogeneity of geology, soils, land use and climate) (Uhlenbrook, 2003). This research therefore focuses on understanding runoff generation processes and applying a hydrological model in a meso-scale catchment. We followed the best practice recommendations for predicting runoff in ungauged basins, suggested by Blöschl *et al.* (2013b). The input data were carefully selected to best represent catchment conditions. Experimental studies were previously conducted in the catchment using tracers (Chapter 6) and baseflow separation techniques (Chapter 5) to understand the runoff generation processes. Furthermore, the approach to derive DRP by Gharari *et al.* (2011), combined with a process-based hydrological model, was used to better understand the key hydrological processes and runoff generation mechanisms in the Kaap catchment.

The specific objectives of this research, in line with the PUB framework, were:

- Interpret key landscape elements with respect to their hydrological functioning, and gather available data for hydrological modelling in the study area;
- Analyse runoff signatures and processes from available gauged catchments;
- Setup a process-based hydrological model that can utilize spatial (gridded) data and that is easy to adapt to different hydrological processes;
- Gradually increase model complexity and assess model sensitivity to different inputs and parameters; and
- Understand the key hydrological processes and runoff generation mechanisms in the catchment, and how this could improve hydrological modelling.

7.2 Materials and methods

7.2.1 Study area

The study area is the same described earlier in Chapters 5 and 6 (sections 5.2.1 and 6.2). Figure 7-1 shows the location, topography, sub-catchments, stations used for this study, as well as land-use, geology and soils. Table 5-1 and Appendix A3 summarizes the catchments' physiographic and hydro-climatic characteristics.

Chapter 3 conducted an extensive analysis of rainfall and streamflow in the Incomati basin, and all the gauges within the Kaap were analysed. The analysis revealed that over the past 60 years (1950-2010), no significant upward or downward trend in the catchment rainfall was found, but rather seasonal variability dominated. The streamflow was analysed using the Indicators of Hydrological Alteration tool, and several significant trends were found in the streamflow records. The Noordkaap gauge (X2H010), for example, showed significant decreasing trends in mean monthly flows, low flows, 7-day minimum flow, among others. Further investigation of this shift in the flow regime identified the change of land use, that is, the increase in forestry plantation, as the main driver of the decreasing trends.

Camacho Suarez *et al.* (2015) conducted an intense tracer study during the rainy season of 2013-2014 in the Kaap catchment. They installed rainfall samplers in two locations in the catchment, and an automatic water sampler at the outlet of the Kaap. Furthermore, grab samples were collected in several locations before rainfall events to provide a snapshot of water quality of the catchment in baseflow conditions. Four major events were sampled and analysed, using isotope and hydro-chemical hydrograph separation, as well as end member mixing analysis. The study revealed great dominance of pre-event water in the streamflow. A three-component hydrograph separation highlighted a major contribution of shallow groundwater, which was enriched with potassium and isotopes. Two main sources of groundwater were identified, the upstream area with fractured granite, characterized by lower ionic content, and the downstream area, with more diverse geology and higher ionic content. Furthermore, a strong correlation was found between antecedent precipitation index and direct runoff. This means that when the catchment is wet from previous rainfall events, and the storages filled, the connectivity of the catchment increases and more direct runoff is generated.

Chapter 5 further explored hydrograph separation in the Kaap, using long-term records of water quality, particularly EC. Baseflow and quickflow components were computed, using a calibrated recursive digital filter, at monthly and annual scales. The digital filter was calibrated using long-term EC and observed flow data. Hydrograph separation showed that all catchments contribute highly to baseflow.

7.2.2 Data used

Hydrological data in the catchment including precipitation, evaporation and streamflow records were collected from the South African Department of Water & Sanitation (DWS, former DWA), the South African Weather Service (SAWS) and the South African Sugarcane Research Institute (SASRI). Figure 7-1A shows the locations of rainfall and streamflow stations, as well as sub-catchment delineation and topography. To analyze the flow behaviour at the outlet and tributaries, average daily discharges at X2H022 (Outlet), X2H008 (Queens), X2H031 and X2H024 (Suidkaap) and X2H010 (Noordkaap) stream gauges were obtained from the DWS. Land use from the Watplan project was used for this analysis (van Eekelen *et al.*, 2015). Topographic information was derived from STRM images with 90 m pixel resolution.

In addition to water use by natural vegetation, the main water users in the catchment are:

- Irrigated sugarcane (98 km^2 area) with a crop water requirement of $92 \cdot 10^6$ m^3 $year^{-1}$ (Mallory and Beater, 2009). However, Mallory and Beater (2009) report that only $62 \cdot 10^6$ m^3 $year^{-1}$ are supplied from the river;
- Domestic water supply to the Umjindi Local Municipality (over 71,200 population), with a demand of $3.9 \cdot 10^6$ m^3 $year^{-1}$ – this is supplied from an interbasin transfer from the neighbouring Lomati dam (part of the Komati catchment) (Mallory and Beater, 2009); and
- Commercial afforestation (considered a streamflow reduction activity) of 443 km^2, with an estimated streamflow reduction of $40 \cdot 10^6$ m^3 $year^{-1}$ (Mallory and Beater, 2009).

There are no major reservoirs in the catchment, and the industrial water requirements are considered insignificant.

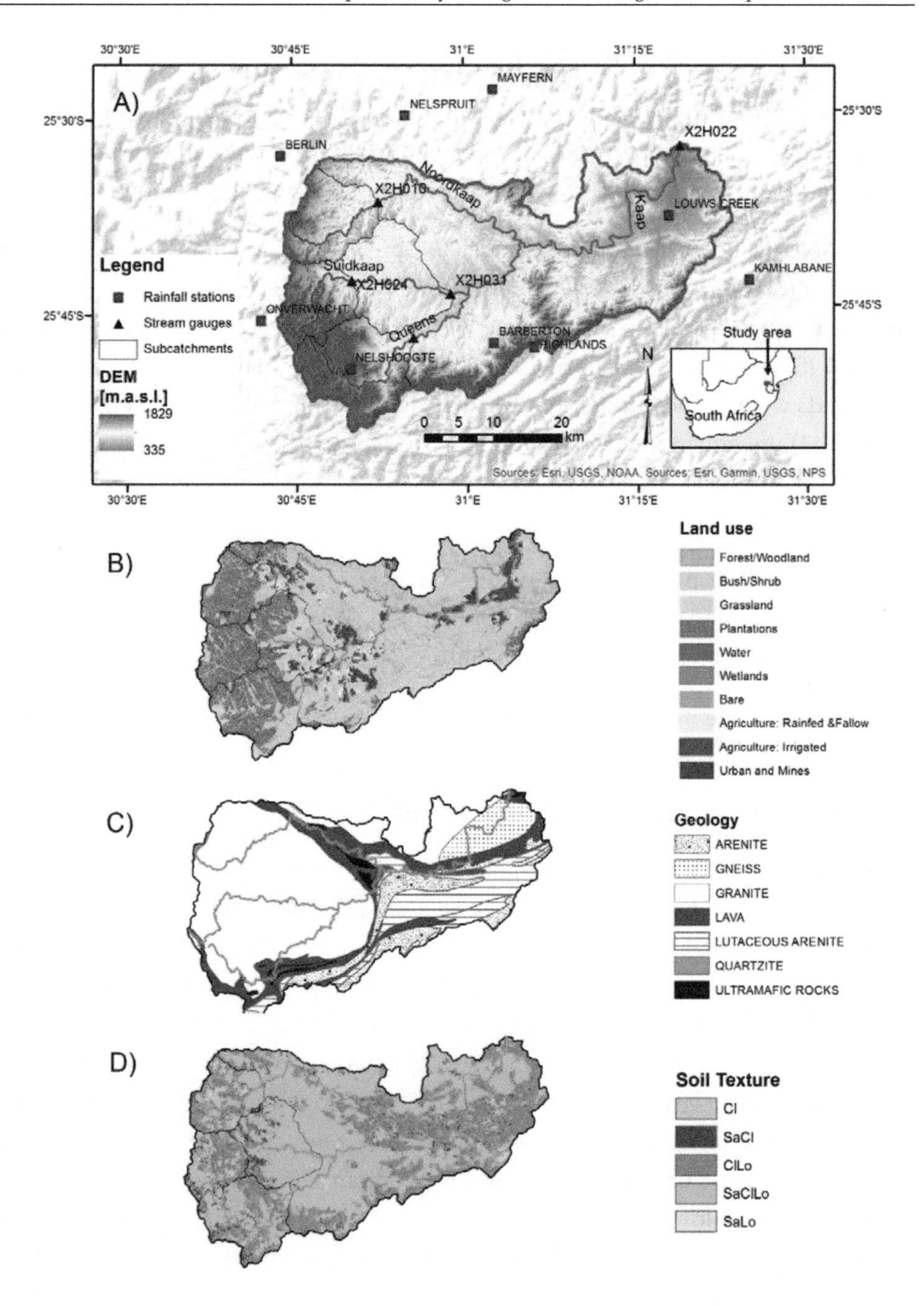

Figure 7-1.A) Location of Kaap catchment in South Africa (inset) and DEM of Kaap catchment with stream gauges, rainfall stations and sub catchment delineation; B) Land-use and land-cover map of Kaap catchment; C) Geological map (Middleton and Bailey, 2009); and (D) Soil texture based on Soil Grids 250m dataset (Hengl *et al.*, 2015; Hengl *et al.*, 2017); the predominantly occurring textures are: Clay (Cl), Sandy clay (SaCl), Clay loam (CL), Sandy clay loam (SaClLo) and Sandy loam (SaLo)

7.2.3 Landscape classification

SRTM images with 90 m pixel resolution were used to define topography. Furthermore, a landscape analysis was conducted to define zones with similar landscape features, which are presumed to have similar runoff generation processes.

The HAND value was computed, as per the procedure of Rennó *et al.* (2008) and Gharari *et al.* (2011). The HAND value was then combined with the slope map, and thresholds were defined to differentiate Wetlands, Hillslopes and Plateaus (or valley bottom) (Savenije, 2010; Gharari *et al.*, 2011) (Figure 7-2A and Appendix A3). The thresholds were defined using expert knowledge and some site verifications. Gharari *et al.* (2011) present an extended calibration procedure to assess sensitivity of HAND model. The thresholds used to define the zones were:

- Stream initiation at 1,000 m.
- The HAND threshold to separate wetlands from Plateau and Hillslope was 10 m.
- The slope threshold to separate Hillslope from Plateau was 12%.

Several runs of the model were conducted and compared with verification locations to adjust the parameters.

7.2.4 Dominant runoff generation zones

After landscape analysis, and in combination with other physiographic information and previous fieldwork (Camacho Suarez *et al.*, 2015), the dominant runoff generation processes were identified in the catchment. The combination of the HAND zones and the geology map helped define zones of slow flow, intermediate (or delayed flow) and fast flow (Figure 7-2B). All wetlands and sealed areas (urban areas and mines) were considered fast flow generation areas. The plateaus had two dominant mechanisms: Plateaus with underlying geology consisting of quartzite and gneiss were classified as intermediate (or delayed) zones, because both vertical and horizontal flows occur. Plateaus with weathered granite and sedimentary rocks were considered slow flow zones because the vertical percolation and recharge to deep groundwater through fissures of the bedrock is the predominant process (Camacho Suarez *et al.*, 2015).

Hillslopes, due to the steep topography, have mostly quickflow occurring through overland flow. When the hillslope has granite, quartzite or gneiss geology, some delayed runoff occurs, as subsurface lateral flow dominates. However, antecedent precipitation can change the dominant processes, in which case, quickflow is sourced from the intermediate runoff zone as well (Camacho Suarez *et al.*, 2015).

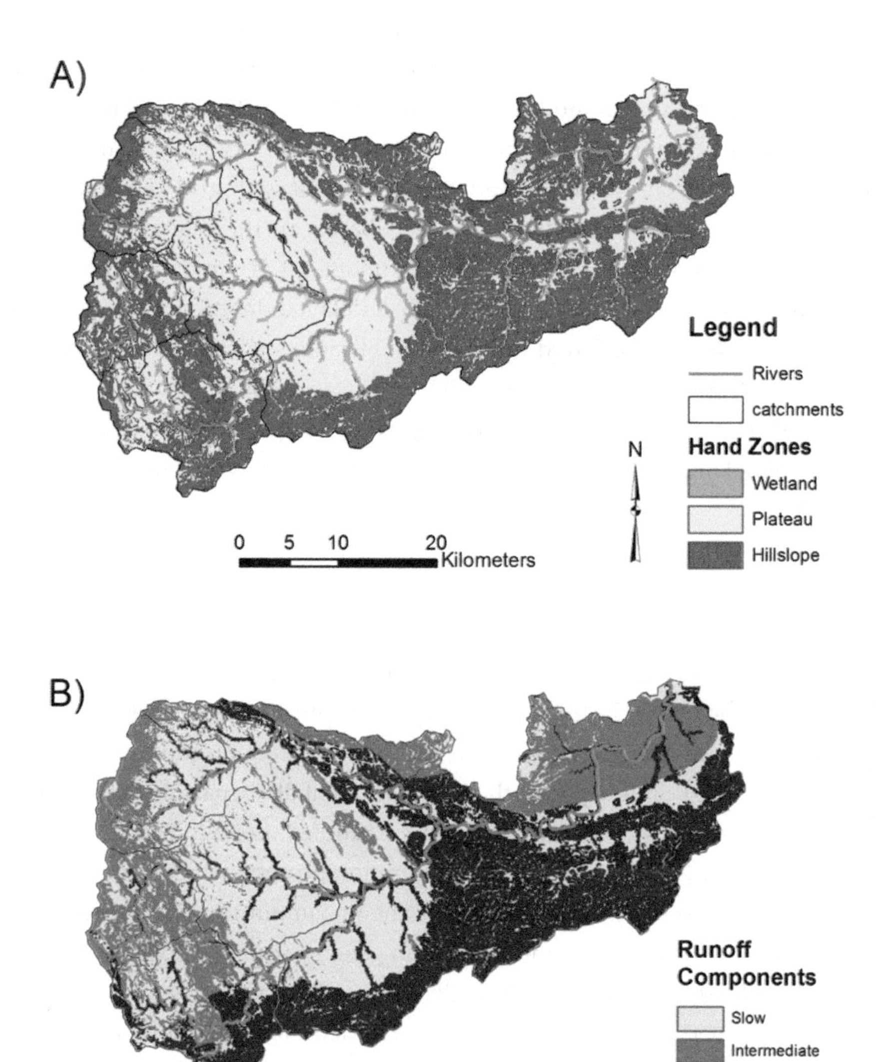

Figure 7-2. A) HAND zones delineated based on combination of HAND value and slope maps; B) Dominant runoff zones, defined based on combination of HAND zones and underlying geology

7.2.5 The STREAM model

The Spatial Tools for River basin Environmental Analysis and Management (STREAM) model (Aerts *et al.*, 1999) has been used in several locations internationally and at different spatial/temporal resolutions. It is a spatially distributed and conceptual model, where the non-linear behaviour of the river basins is explained by a combination of thresholds and linear reservoirs. The model is based on a raster GIS which calculates the water balance of each grid cell and routes this through a stream channel network which is based on the digital elevation model (DEM). There is no routing of the surface runoff – it is removed from the model within the same time step as it is generated. A detailed description of model genesis and configuration can be found in several publications (Gerrits, 2005; Winsemius *et al.*, 2006; Kiptala *et al.*, 2014). The model was selected because of its ability to use distributed (raster) data, and ease of configuration in open source PCRaster dynamic programming language. The main model parameters and variables are presented in Table 7-1.

The model was used as a tool to test our process understanding in the studied catchments and to highlight shortcomings in process representation in the model. The model structure included some of the main processes expected in a semi-arid catchment such as precipitation, interception, evaporation, and runoff generation (Figure 7-3). After interception, the effective rainfall is partitioned using the *cr* coefficient between the unsaturated and saturated zones. The portion in the unsaturated zone is available for the transpiration process, which is computed using the soil water balance and is regulated by the maximum unsaturated zone storage, *Sumax*. The portion in the saturated zone can generate runoff, if certain groundwater storage thresholds are exceeded. Initially, the capillary rise process, whereby water from the saturated zone returns to the unsaturated zone was not simulated, but in later model runs this process was also. Previous research showed that plantation forest (*Eucalyptus*) can access water from great depths, depleting the groundwater storage (Dye, 1996; Scott and Lesch, 1997; Jewitt, 2006b). To mimic this process, the *Sumax* parameter, which was defined based on rooting depth of dominant land use and available water content (based on soil hydraulic properties), was made much large under forest and plantation land-uses. Where shallow vegetation predominates, the *Sumax* parameter was set at low values. Note that the hydrological model did not consider direct abstractions of water for irrigation and other purposes.

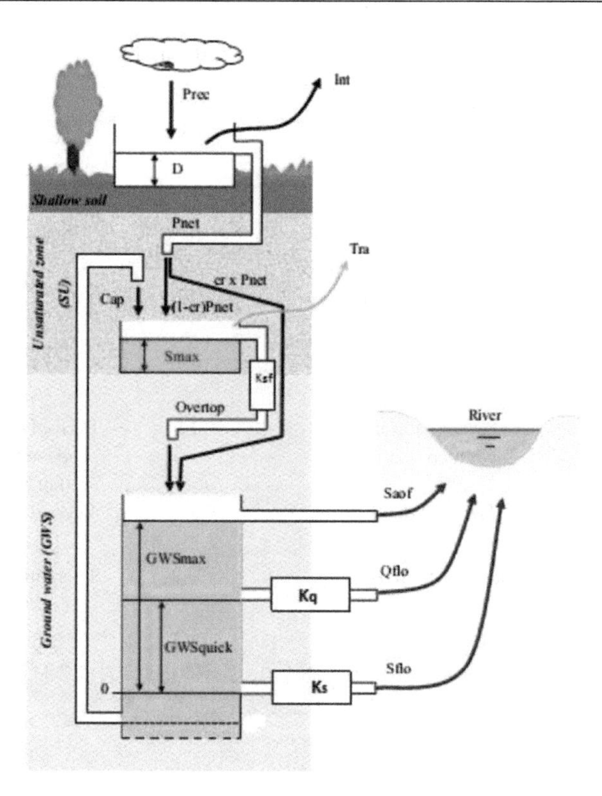

Figure 7-3. Model configuration (Gerrits, 2005; Winsemius *et al.*, 2006)

7.2.6 Model inputs, parameters and setup

The hydrological model was configured to simulate stream flow for the period 2003-2013 for the Kaap catchment, with a daily time step and 90m cell grid size.

Daily rainfall station data were used for precipitation. The data were interpolated with the Inverse Distance Weighing method (IDW) and were corrected with an elevation factor derived from the Mean Annual Precipitation map (Schulze *et al.*, 2007), according to the methodology described in Sieber and Uhlenbrook (2005). Interception was defined using a fixed daily threshold coefficient *D* based on the land use and land cover map, listed in Table 7-2. Potential Evaporation was also derived from the station data, and interpolated using IDW method. Actual Transpiration was computed from the soil moisture water balance in the unsaturated zone. The *Sumax* parameter was derived from a combination of available water content (field capacity minus wilting point of each soil type) and rooting depth of each respective land cover. All model parameters were derived from careful analysis of the literature, local expert knowledge and by calibration, as explained in Table 7-1.

Table 7-1. Model parameters and variables

Parameter	Unit	Description	Value	Estimation method
Ku	day	Overtop timescale	0.5	Recession curve Analysis
Ksf	day	Saturation overland flow timescale	1	Recession curve Analysis
Kq	day	Quickflow timescale	5	Recession curve Analysis
Ks	day	Slow flow (baseflow) timescale	100	Recession curve Analysis
Suini	mm	Initial storage in unsaturated zone	20	Calibration
GWSini	mm	Initial ground water storage in saturated zone	20	Calibration
LP	-	Reduction of potential evapotranspiration	0.5	Literature
Cr	-	Unsaturated/saturated zone separation coefficient	0-1	Derived from slope, soil texture and land-use land-cover map (Liu and De Smedt, 2004)
Zr	m	Rooting depth	0.5-2.5	Literature/Land-use land-cover map
D	mm/d	Interception threshold	0 - 4	Literature/Land-use land-cover map
Qc	-	Quickflow coefficient	0 - 1	Calibration/Soil Texture
Sumax	mm	Maximum storage in unsaturated zone	0-500	Field capacity and rooting depth
GWSmax	mm	Maximum ground water storage in saturated zone	25lnGWSdem	(Gerrits, 2005)
Cflux	mm/d	Maximum capillary rise threshold	0 - 2	(Kiptala *et al.*, 2014)
GWSmin	mm	Minimum ground water storage threshold to initiate capillary rise	(0-0.5)**GWSmax*	Modified after (Kiptala *et al.*, 2014)

7.2.7 Model simulations

Several model configurations with stepwise variation of model inputs, parameters and processes of differing complexity were tested. The following were the main simulation comparisons conducted:

- Rainfall input (Station data with Thiessen regionalization, with Inverse Distance Weighing and elevation correction, and Remote sensing precipitation from Chirps database).
- Unsaturated/saturated zone separation coefficient *Cr* [-].
- Maximum ground water storage in saturated zone parameter *GWSmax* [mm], derived from DEM or from HAND maps.
- Implementation of capillary rise process, with different thresholds of *Cflux* [mm/d].

- Combinations of capillary rise and different *cr* parameters.
- Implementation of capillary rise with initiation threshold *GWSmin* (*GWSmin* =[0.5, 0.2, 0.1, 0.01] * *GWSmax*).
- Maximum storage in unsaturated zone parameter *Sumax* [mm].

In addition, the HBV model (Bergström, 1992; Lindström *et al.*, 1997) was set up for the catchments for comparison. The model was configured using similar input data (precipitation, temperature and potential evaporation), but only vegetation and elevation band zones were used to discretize the model. Automatic calibration was applied to obtain the best performing parameter sets.

7.2.8 Runoff signatures and assessment of model performance

The Prediction in Ungauged Basins (PUB) book (Blöschl *et al.*, 2013b) suggests a framework for hydrological understanding of catchments, by focusing on their runoff signatures. There are a myriad of possible signatures, but we choose to focus on the key signatures suggested by Blöschl *et al.* (2013b), which are commonly used in the region as well (e.g. Mazvimavi *et al.*, 2005): annual runoff, seasonal runoff, flow duration curve (FDC), low flows, floods, and runoff hydrographs.

The model performance was also assessed visually and statistically using different indicators of goodness of fit of the hydrographs: the Nash-Sutcliffe efficiency (NSE), the Logarithmic Nash-Sutcliffe efficiency (LogNSE), Bias and percentage Bias (PBias), Mean absolute error (MAE), the Pearson R^2, Root mean square error (RMSE), and Kling-Gupta efficiency (KGE) coefficient (Gupta *et al.*, 2009; Kling *et al.*, 2012). The NSE varies between $-\infty$ and 1.0, with 1.0 being the optimal value. Values between 0 and 1 are considered acceptable, whereas less than 0 is unacceptable performance. LogNSE has a similar range, but the flow values are transformed into logarithmic to better analyse low flows. Bias, MAE and RMSE have the same unit as observed flow, whereas PBias is the percentage of bias in relation to mean flow; the closer to 0, the better the model performance, with low-magnitude values indicating accurate model simulation. Positive values indicate model underestimation bias, and negative values indicate model overestimation bias. R^2 varies between 0 and 1, whereas KGE varies between and $-\infty$ and 1. In both cases values between 0.7 and 1 are considered good; between 0.5 and 0.7, acceptable; and below 0.5, poor. The KGE also offers diagnostic insights into the model performance because of the decomposition into correlation, bias term and variability term. From a hydrologic perspective usage of KGE assists in reproducing temporal dynamics, as well as preserving the distribution of flows; therefore, this was adopted as the main indicator of goodness of fit.

7.3 Results

7.3.1 Model parameterization

The final STREAM model parameters used for comparing simulations are listed in Table 7-2 and Table 7-3, and illustrated in Figure 7-4. Several manual calibration runs were conducted, where each parameter was varied while others kept constant. The best performing parameter sets were retained for subsequent simulations.

For the Sumax parameter, the areas with forest and plantations had higher Sumax values, because these occur on locations with deeper soils and stronger baseflow, indicative of larger water storage; research has shown that these vegetation types can tap deep water stores.

The cr parameter was derived from a combination of land use, soil texture and slope (Liu and De Smedt, 2004). The qc parameter, however, was mainly driven by soil type. Coarser soils, such as sandy loams or sandy clays have a lower qc threshold, because these soils allow for quicker response, the threshold to initiate quickflow being lower. The finer clayey soils, in contrast, can hold water for longer periods of time, increasing the qc threshold value. The GWSmax parameter followed closely the elevation pattern, and the relationship derived by Gerrits (2005) was used.

Table 7-2. Final model parameters dependent on land use and land cover map.

Land use and Land cover	%Total area	*D* [mm/day]*	*Zr* [m]*
Forest/Woodland	19.9%	4	2.5
Bush/Shrub	31.6%	2	1
Grassland	13.9%	2	0.8
Plantations	23.3%	4	2.5
Water	0.2%	0	0.5
Wetlands	0.7%	1	0.5
Bare	0.3%	1	0.5
Agriculture: Rain-fed, Planted pasture, Fallow	2.6%	2	1.5
Agriculture: Irrigated	5.8%	2	2
Urban and Mines	1.7%	1	0.5

* *D* is the interception threshold and *Zr* is the rooting depth, used to compute the *Sumax* parameter

Table 7-3. Quickflow separation coefficient, based on soil texture

Soil Texture	*Qc* [-]
Clay (Cl)	0.9
Sandy clay (SaCl)	0.7
Clay loam (CL)	0.8
Sandy clay loam (SaClLo)	0.6
Sandy loam (SaLo)	0.5

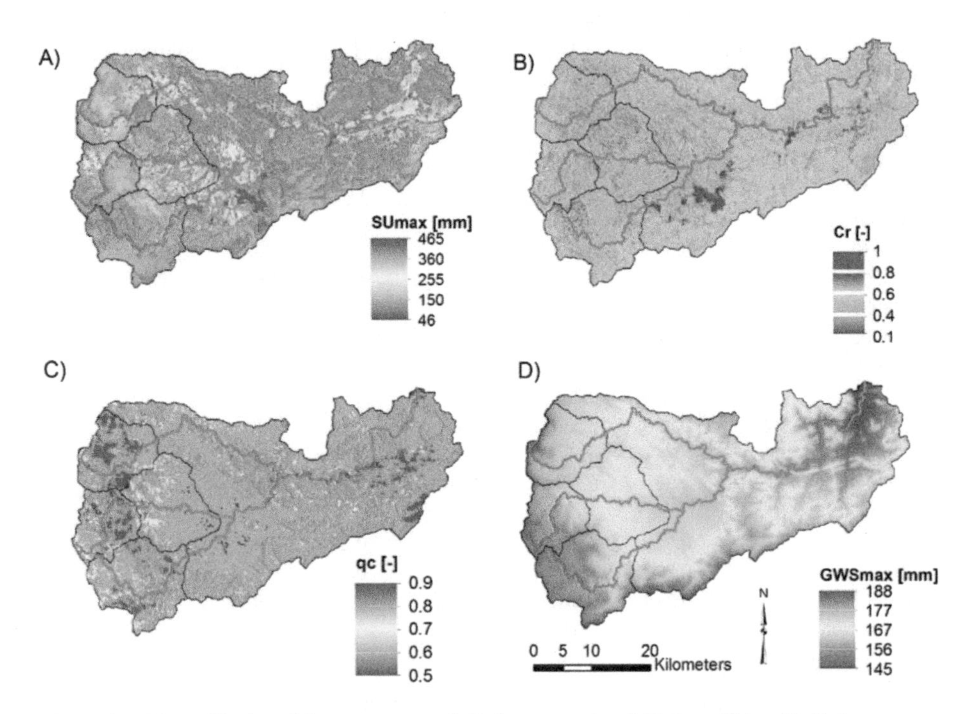

Figure 7-4. Maps of final model parameters used. A) *Sumax* map [mm]; B) *Cr* coefficient [-]; C) *Qc* coefficient [-]; D) *GWSmax* [mm]

7.3.2 Model simulations

Over 70 model runs were conducted, but only a sample of four representative runs will be presented and discussed (Table 7-4). Overall, the model simulations were able to capture the flow dynamics well. However, in several runs, the model overestimated peak flows and baseflows, especially at the Kaap outlet and in Queens.

Table 7-4. Characteristics of model runs presented in the results section

Run	*Cflux* [mm/d]	GWSmin	Description
53	0	0	Model without capillary rise implemented
60	1	0	Model with capillary rise implemented but no *GWSmin*
64	2	0.1*GWSmax*	Model with capillary rise implemented and *GWSmin*
67	2	0.01*GWSmax*	Model with capillary rise implemented and *GWSmin*

Table 7-5. Goodness of fit indicators for the selected model runs, at monthly time scale.

	Bias [m³/month]	KGE [-]	LogNSE [-]	MAE [m³/month]	NSE [-]	PBias [%]	RMSE [m³/month]	Pearson R^2 [-]
Run53								
X2H010	22.8	-0.29	-0.77	22.8	-2.46	170.9	28.1	0.83
X2H008	48.4	-1.23	-1.26	48.4	-3.97	479.1	59.2	0.83
X2H031	52.1	-0.91	-1.66	52.6	-5.72	269.3	64.3	0.76
X2H022	371.7	-2.76	-1.38	371.7	-11.44	7299.9	424.5	0.84
Run60[a]								
X2H010	-0.4	0.68	na	8.1	0.53	-16.6	10.4	0.84
X2H008	14.4	0.32	na	18.8	-0.13	88.9	28.2	0.82
X2H031	5.3	0.62	na	17.8	0.05	8.6	24.1	0.76
X2H022	85.6	0.12	na	102.2	-0.46	186.9	145.3	0.84
Run64								
X2H010	-4.2	0.67	0.38	6.7	0.52	-15.4	10.5	0.77
X2H008	7.2	0.61	0.53	12.1	0.50	87.4	18.7	0.83
X2H031	-0.8	0.75	0.58	10.6	0.52	10.6	17.1	0.75
X2H022	57.7	0.32	0.10	72.4	0.42	1829.5	91.6	0.84
Run67								
X2H010	-8.4	0.22	-6.10	9.6	0.30	-58.9	12.6	0.79
X2H008	1.4	0.79	-0.44	11.3	0.55	-2.3	17.7	0.82
X2H031	-9.8	0.34	-4.14	13.7	0.37	-49.1	19.7	0.76
X2H022	0.7	0.83	0.35	47.5	0.66	197.7	70.6	0.84

[a] LogNSE could not be computed for run 60 because there were months with zero flow simulated.

Note: KGE is the Modified Kling Gupta Efficiency, LogNSE is the Logarithmic Nash Sutcliffe efficiency coefficient, MAE is the Mean Absolute Error, NSE is the Nash Sutcliffe efficiency coefficient, RMSE is the Root Mean Squared Error and PBias is the Percent Bias.

A comparison of the goodness of fit indicators was done to see which model better represents the actual catchment conditions (Table 7-5). Overall, the Pearson correlation was good, ranging from 0.75 to 0.84, meaning that the simulated flows generally followed well observed flow pattern. The NSE was poor to acceptable, mostly due to the overestimation of flow in some runs (eg. run 53 and run 60), and the seasonality of the flow. In the best performing runs, the NSE was 0.5 to 0.66. Run 60 simulated zero flows during the low flow season, and thus it was not possible to calculate LogNSE. In terms of KGE, which is the most integrated indicator, run 64 was the best for Noordkaap and Suidkaap catchments, with KGE of 0.67 and 0.75, respectively, whereas run 67 was the best for the Queens and Kaap catchments, with KGE of 0.79 and 0.83, respectively. The Bias was very high, especially in the Kaap outlet. These results show how different model setups are needed for the different catchments. However, there is a similarity between Noordkaap and Suidkaap, and also between the Queens and the Kaap.

Furthermore, a visual analysis of the different hydrological signatures was conducted to further understand which processes were better represented by each model setup. From the other model simulations (not reported here), the parameters *cr* and *qc* proved to be very sensitive.

7.3.3 Comparison of runoff signatures

7.3.3.1 Annual runoff

The annual runoff, which is a key component of the water balance, was computed for all hydrological years. There was difficulty in closing the water balance in the initial runs, when a simple setup without capillary rise (or feedback from the groundwater storage to the unsaturated zone) was implemented. Figure 7-5 illustrates the results of run 64, compared to observed flow for the four catchments.

Regarding the annual dynamics, one can see that the model tends to better capture the flows generated in wetter years than in drier years. This may be due to more uncertainty in the storage conditions during drier years, and the impact of water abstractions for irrigation. The naturalized flow (Bailey and Pitman, 2015) is also 17 to 51% higher than observed flow, which implies that water abstractions and reductions in streamflow could be up to 50%, particularly in the dry season.

One important aspect to consider is that in semi-arid and sub-humid areas, the evaporation component of the water balance is very large. Actually, the potential evaporation is much larger than rainfall, which reflects in more than 90% of the water balance being attributed to evaporation, and only 10% or less to runoff generation (Table 7-6). Therefore, uncertainties related to the computation of evaporation, such as the parameters used, interception, and the interpolation of input data, can greatly affect the results of model simulations. This is illustrated by the great difference between evaporation estimates from different model runs (Table 7-6). A comparison was also made between evaporation generated by the water balance model, and evaporation from remote sensing products (Table 7-6). A more detailed description of the remote sensing evaporation products is available in the supplementary material (Appendix A4). Comparing monthly and annual scales revealed that both ALEXI (Anderson *et al.*, 1997; Hain *et al.*, 2009) and CMERST (Guerschman *et al.*, 2009) products generally overestimate actual evaporation, whereas SSEBop (Senay *et al.*, 2013; Chen *et al.*, 2016) results in an underestimate.

Table 7-6. Average annual water balance, including evaporation from remote sensing products for comparison, for the period 2003-2013.

	Noordkaap X2H010 Mean ± Stdev	**Queens X2H008** Mean ± Stdev	**Suidkaap X2H031** Mean ± Stdev	**Kaap X2H022** Mean ± Stdev
Rainfall [mm/year]	1008.0 ± 154.1	1126.6 ± 181.6	946.3 ± 135.9	774.0 ± 121.6
FlowObs [mm/year]	137.5 ± 66.7	127.4 ± 84.5	106.8 ± 55.0	61.4 ± 41.1
RC [%]	13% ± 5%	11% ± 0.1	11% ± 0.0	7% ± 4%
Qnat [mm/year] [a]	222.0 ± 64.0	153.0 ± 70.0	217.0 ± 49.0	106.0 ± 36.0
Fm53 [mm/year]	335.7 ± 67.5	406.4 ± 74.4	316.0 ± 53.9	298.1 ± 48.2
Fm60 [mm/year]	144.7 ± 61.9	210.4 ± 75.8	131.0 ± 49.4	117.2 ± 40.5
Fm64 [mm/year]	113.9 ± 40.5	169.1 ± 59.7	106.9 ± 33.1	99.6 ± 25.6
Fm67 [mm/year]	78.9 ± 45.3	135.7 ± 64.2	71.3 ± 36.4	63.6 ± 28.1
PminFlowOb [mm/year]	870.5 ± 112.8	999.2 ± 121.5	839.5 ± 94.7	712.7 ± 85.8
ETfao [mm/year]	1250.7 ± 69.2	1221.3 ± 93.6	1225.0 ± 89.5	1235.3 ± 83.9
ETm53 [mm/year]	672.0 ± 93.7	717.9 ± 109.3	660.1 ± 88.5	503.7 ± 72.5
ETm60 [mm/year]	864.2 ± 98.0	915.3 ± 109.8	838.2 ± 93.3	681.7 ± 83.8
ETm64 [mm/year]	894.7 ± 120.0	956.4 ± 126.1	853.7 ± 108.3	691.9 ± 99.3
ETm67 [mm/year]	930.0 ± 114.8	990.1 ± 121.4	890.1 ± 106.1	728.9 ± 97.7
ETal [mm/year]	1100.9 ± 54.3	1079.6 ± 55.3	884.7 ± 41.7	861.4 ± 35.8
ETcm [mm/year]	1127.4 ± 269.8	1142.6 ± 56.3	1005.0 ± 50.9	831.2 ± 181.3
ETss [mm/year]	788.5 ± 35.9	733.8 ± 59.8	608.2 ± 68.4	608.4 ± 53.9

[a] Qnat is the average for 2003-2010, given that naturalized flows are only available up to 2010 hydrological year.

Note: FlowObsv is observed flow, RC is runoff coefficient, Qnat is the naturalized flow obtained from WR2012 database, Fm53 to 67 are the simulated flows for runs 53 to 67, PminFlowOb is the difference between precipitation and observed flow, ETfao is the potential evaporation using FAO method, ETm53 to 67 are the simulated actual evaporation for model runs 53 to 67, ETal is evaporation from ALEXI product, ETcm is evaporation from the CMERST product, and ETss is evaporation from the SSEBop product.

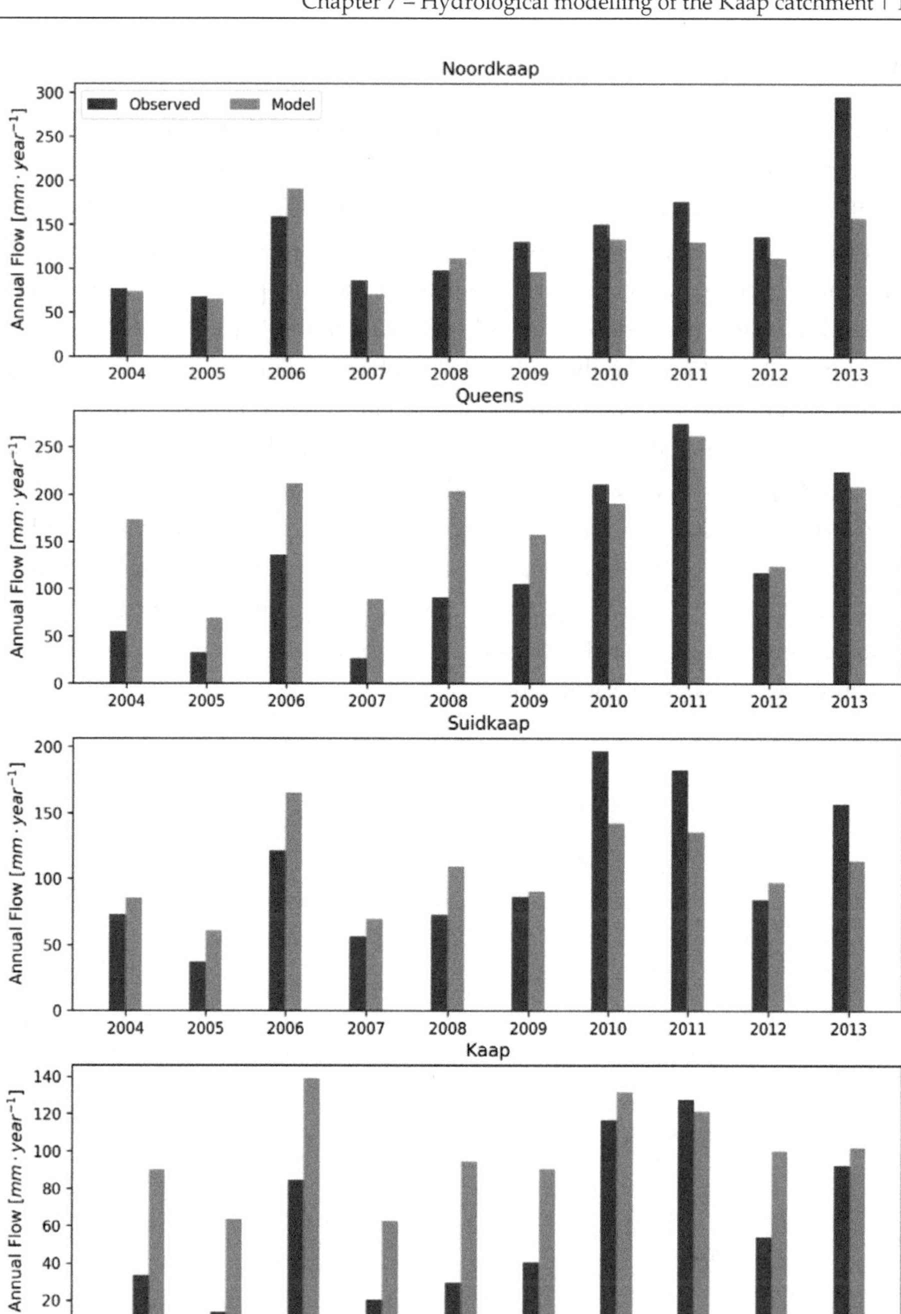

Figure 7-5. Annual Flow for the four catchments, using results of run 64

7.3.3.2 Seasonal runoff

The average monthly streamflow graphs (Figure 7-6) show the strong seasonality of streamflow in the catchments. The flow components analysis demonstrates that quickflow and saturated overland flow are only active for few months of the year (November to March, with peak contributions in January/February). For the Noordkaap catchment, model run 64, which included capillary rise and *GWSmin* threshold, was able to capture the monthly flow pattern relatively well. This pattern was also well captured in the Suidkaap catchment. For the Queens and Kaap catchments, however, this model run overestimates flow, especially during the wet months. Another model configuration (model 67, available in the supplementary material, Appendix A4), which included a higher coefficient for capillary rise, generated better results for these latter two catchments. Overall, from the results of run 64, the baseflow component accounted for 85% of the flow in the Queens catchment, 95% in the Nordkaap, 94% in the Suidkaap, and 93% in the Kaap. The quickflow contribution ranged between 4 and 13%, whereas saturated overland flow was about 1% or less for all catchments.

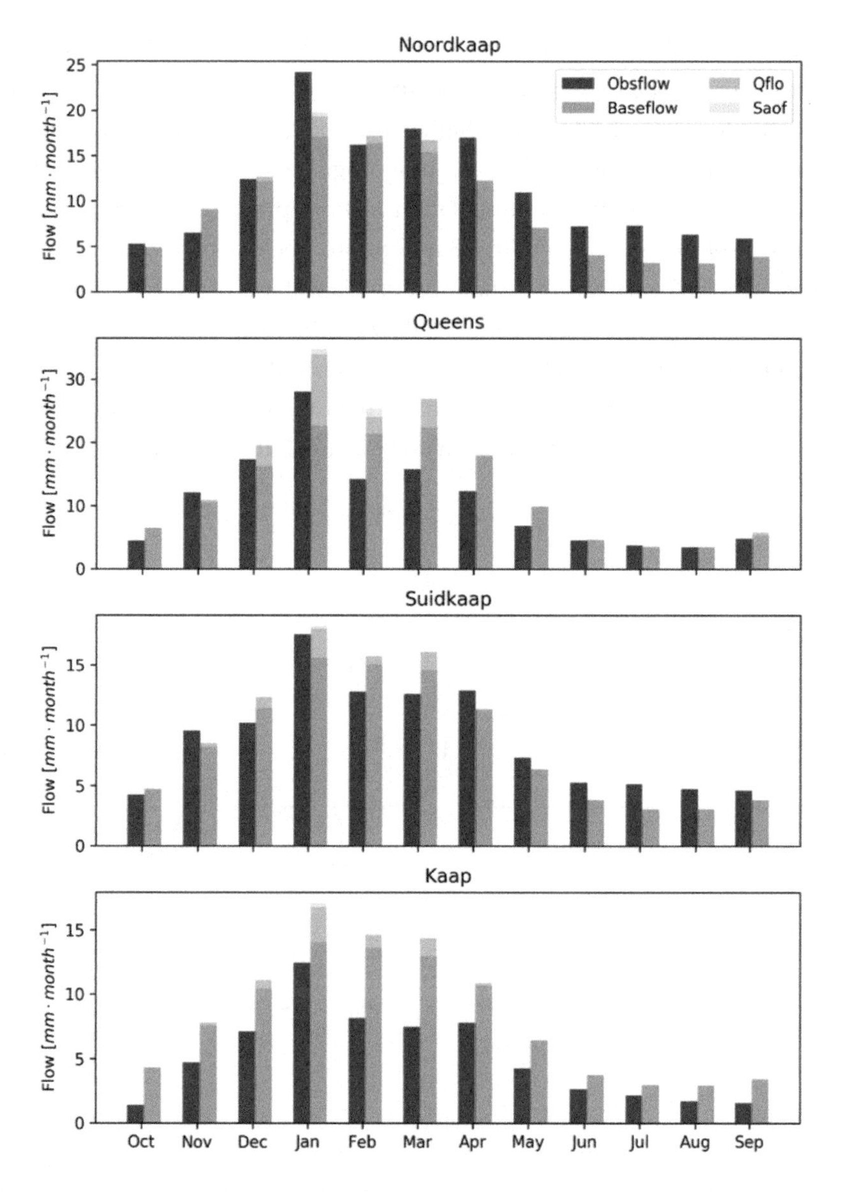

Figure 7-6. Mean monthly observed and modelled flow for the four catchments, using results of run 64. Modelled flow is partitioned in the components: baseflow, saturated overland flow (Saof) and quickflow component (Qflo)

7.3.3.3 Flow duration curves

The flow duration curves (Figure 7-7) give a comprehensive overview of the flow regime. Once more it is evident that the flow variation for the Noordkaap and Suidkaap catchments is fairly well represented. However, for the Queens and Kaap catchments the model overestimates the middle and low flows. This overestimation could be because of the representation of subsurface flow processes in the catchment model. Both Queens and Kaap have higher percentages of area with hillslopes, and also more diverse geology and soils, including a variety of sedimentary rocks. Apparently a more complex representation of flow processes and groundwater is required to capture such variability. Furthermore, the Kaap catchment has higher water abstractions for irrigation – and most of this water is abstracted during the dry season and in drier years. The naturalized flow for the Kaap catchment is 42 % higher than the observed flow, which can largely be attributed to water abstractions for irrigation. Most of the irrigated sugarcane in the catchment is located in the Lower Kaap valley, with an estimated irrigation demand of $92 \cdot 10^6$ m^3/year.

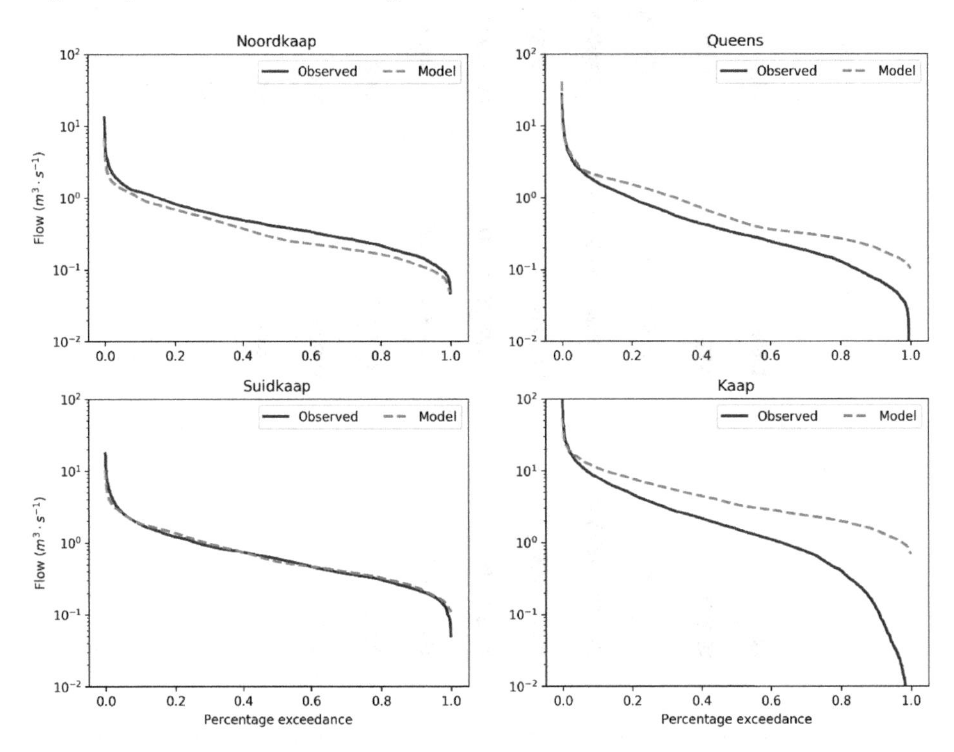

Figure 7-7. Flow duration curve for the four catchments, using results of run 64.

7.3.3.4 Low flows and floods

The low flows and floods can be characterized by their magnitude, frequency and duration. The flow duration curve provides commonly used indicators of low flows and high flow regime. Q95 (flow exceeded 95% of the time) is the most frequently used indicator of low flow. Q75 is frequently used in South Africa for yield estimation, and the Q50 and Qmean are also reported. On the high flow side, Q1 and Q5 are used, as well as floods with 100-year recurrence intervals. However, given that the model dataset is only 10 years long, it was not sufficiently long to perform a flood frequency analysis. For the sake of comparing the model runs, Q1 and Q5 for modelled and observed time series were instead compared as indicators of high flows. Table 7-7 compares flow percentiles for observed and modelled streamflow under different model setups. The slope of the flow duration curve, computed as the slope between Q30 and Q70, is also reported for comparison. This set of signatures reveals once more the difficulty of having one single model setup performing equally well for both high and low flows, while getting the same slope of the FDC. This is likely due to the fact that in these catchments, different processes control the low flow and high flow generation; apparently the simple model structure did not fully capture these differences.

Table 7-7. Quantiles of high and low flow, and slope of flow duration curve (between Q30 and Q70) for the different model runs compared to observed flow (FlowObs)

		Q01	**Q05**	**Q50**	**Q75**	**Q90**	**Q95**	**Q99**	**Qmean**	**SlopeFDC**
		m^3/s	m^3/s	m^3/s	m^3/s	m^3/s	m^3/s	m^3/s	m^3/s	[-]
Noordkaap	FlowObs	3.12	1.60	0.40	0.24	0.16	0.12	0.09	0.59	1.48
	Fm53	5.68	3.03	1.12	0.66	0.47	0.41	0.30	1.34	1.51
	Fm60	2.96	1.68	0.38	0.02	0.00	0.00	0.00	0.58	3.33
	Fm64	2.04	1.32	0.27	0.18	0.12	0.10	0.07	0.46	1.74
	Fm67	1.92	1.23	0.10	0.01	0.01	0.01	0.01	0.32	2.97
Queens	FlowObs	5.94	2.49	0.32	0.16	0.07	0.05	0.03	0.73	1.55
	Fm53	12.27	6.49	1.76	1.03	0.75	0.64	0.49	2.32	1.32
	Fm60	8.11	3.49	0.76	0.09	0.00	0.00	0.00	1.20	2.98
	Fm64	5.89	2.69	0.48	0.29	0.20	0.16	0.12	0.96	2.01
	Fm67	5.54	2.58	0.26	0.02	0.02	0.01	0.01	0.77	3.14
Suidkaap	FlowObs	5.71	2.57	0.60	0.34	0.23	0.18	0.11	0.91	1.44
	Fm53	12.21	6.22	2.18	1.26	0.91	0.78	0.58	2.62	1.50
	Fm60	5.93	3.19	0.69	0.04	0.00	0.00	0.00	1.09	3.29
	Fm64	4.05	2.56	0.55	0.36	0.24	0.19	0.13	0.89	1.62
	Fm67	3.75	2.34	0.17	0.03	0.02	0.02	0.01	0.59	2.79
Kaap	FlowObs	25.06	12.20	1.55	0.57	0.12	0.04	0.00	3.28	1.72
	Fm53	65.57	36.15	13.10	7.52	5.47	4.74	3.52	15.49	1.53
	Fm60	32.04	18.76	3.76	0.30	0.00	0.00	0.00	6.09	3.21
	Fm64	22.61	14.67	3.35	2.18	1.48	1.14	0.87	5.18	1.62
	Fm67	20.46	13.06	1.09	0.15	0.11	0.09	0.06	3.30	2.77

7.3.3.5 Monthly hydrographs

Figure 7-8 shows the hydrographs for catchments aggregated at monthly time scale for the entire simulation period. During a sequence of wet years (2010 to 2013), the model was able to represent the flow dynamics fairly well with no systematic under- or over-prediction of flows. In drier years, better characterization of the evaporation processes is required, as well as groundwater discharge, storage and water abstractions, as these greatly influence the water balance. Small, localised rainfall events are also very difficult to capture with the given monitoring network. Water abstractions for irrigation and other uses are relatively higher in drier years, so explicit representation of the irrigation management and other water uses would be required to improve the model.

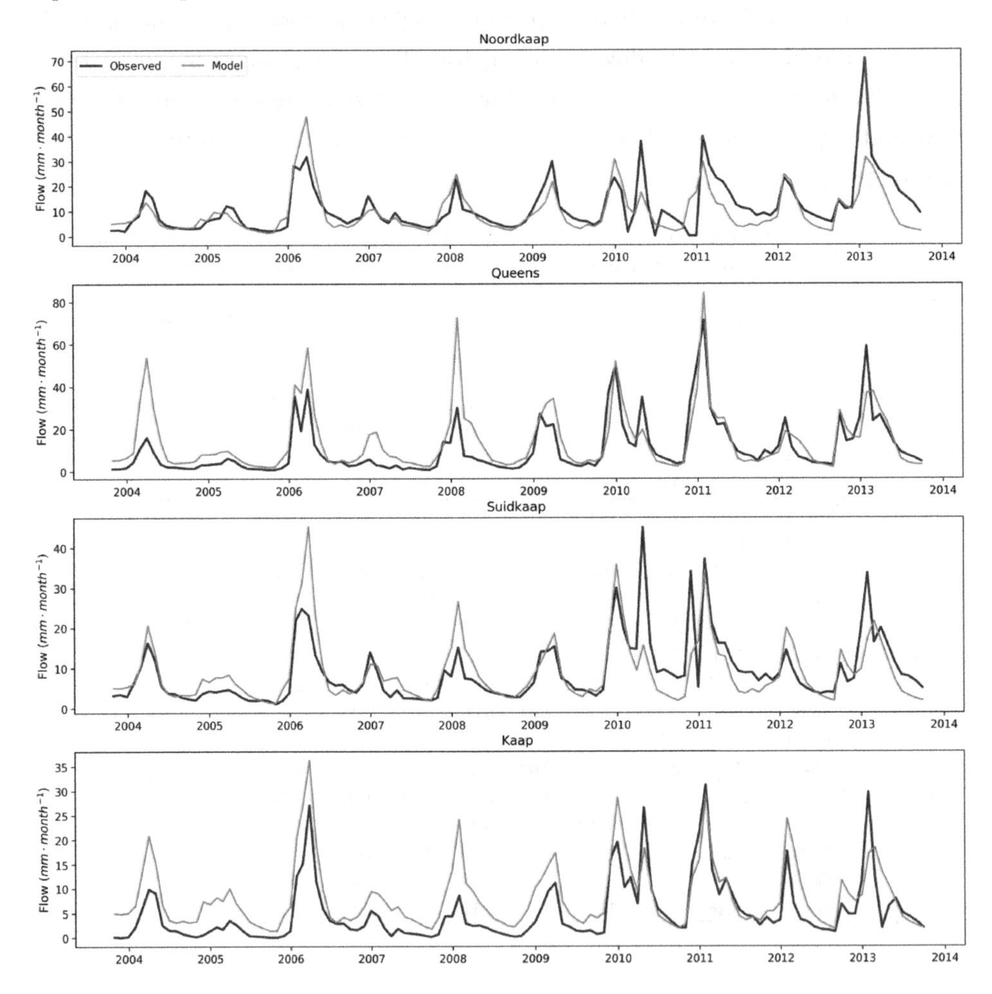

Figure 7-8. Monthly hydrographs for the four catchments, using results of run 64.

7.4 Discussion

7.4.1 Implications for hydrological process understanding

The research provided a framework to evaluate model performance considering a range of information sources. The use of landscape features to map dominant runoff mechanisms also assisted in estimating parameters for the STREAM model. Model parameter estimates derived from this study could guide parameterisation in the case of other semi-arid catchments with similar characteristics. Furthermore, the research highlighted gaps in understanding of key hydrological processes in the study catchments, which seem typical for semi-arid areas. The computation of all evaporation fluxes and the accurate quantification of water uses is very important for the water balance estimates of semi-arid areas, given that the greatest component of the water balance is attributed to these processes. The use of information from previous hydrograph separation studies (using digital filters or water quality data) proved useful to aid in understanding the flow components and dominant flow generation processes in the catchments as well. Therefore, it is relevant for other areas to explore these types of data and information, in addition to traditional hydrological data.

Notably, the soil grids dataset that is freely available at global scale proved useful as model input.

7.4.2 Implications for water resource management

The results reveal that there are great heterogeneities in the catchments studied. Water resource management decisions are made at the scale of the catchments studied; therefore, it is important that a better process understanding is included in current management models and tools.

We could see that especially the Queens and Kaap catchment seem to have higher levels groundwater/surface water interaction. These catchments are more impacted by water abstractions and evaporation, particularly during dry seasons and drier years, which is evident from the steeper shape of the flow duration curves in the low flow portion (Figure 7-7).

The Noordkaap and Suidkaap catchments, in contrast, are dominated by subsurface runoff, as a result of deeper soils and mostly fractured granite lithology. The baseflow is larger in these catchments, and they are likely the recharge areas of the regional groundwater body. Research using hydro-chemical and stable isotopes in the Kaap outlet over the rainy season of 2013-2014 (Camacho Suarez *et al.*, 2015) revealed that 64 to 98% of flow in the Kaap was from shallow and deep groundwater components. During wet conditions, up to 41% of total runoff was attributed to direct runoff, and strong correlations were found between antecedent precipitation conditions and direct runoff.

Saraiva Okello *et al.* (2018b) (Chapter 5) also found high contributions of baseflow to total flow in the Kaap catchment and tributaries using calibrated recursive digital filters for hydrograph separation. They reported baseflow contributions ranging from 45 to 70%, with very high inter and intra annual variability.

Water abstractions for irrigated agriculture significantly impact the streamflow in the Kaap, particularly in the dry season and in drier years. This impact seems to be more pronounced at the Kaap outlet, and in the Queens catchment, and is attributed to the different processes that are dominant in these catchments. The Suidkaap catchment, in spite of also experiencing high streamflow reductions (mostly due to commercial forestation), still sustains significant baseflow contributions. Some water is imported to augment irrigation in the Kaap valley, but these water imports do not significantly affect the water resources of the Kaap catchment, as return flows are small.

The reduction of water quantity in the Kaap negatively affects the water quality in the catchment. Saraiva Okello *et al.* (2018b) (Chapter 5) reported that higher loads of EC and other water quality parameters, due to reduced dilution capacity of the system, particularly during the dry season and in dry years. Deksissa *et al.* (2003) and Slaughter and Hughes (2013) supported this finding, attributing it to the combination of abandoned mines and irrigation return flows that occur in the Kaap. Therefore, in order to improve water quality in the Kaap and in the Crocodile catchment overall, it is important to reduce water abstractions and/or better control and restrict water pollution in the sources.

7.4.3 Input uncertainty and model structure

The rainfall input greatly impacts the runoff generation. The regionalization of rainfall based on available stations using the IDW method could have induced a greater number of rainy days and reduced the magnitude of rain events. This results in higher interception rates, and underestimation of flow peaks. An attempt was made to compare the rainfall regionalization with the Thiessen method, and with remote sensing rainfall (Chirps), but still the IDW method provided better simulation results, likely due to the altitude correction using the MAP pattern.

Currently, the model only accounts for human activity in terms of modified land use, and not through explicit water abstractions. Improved monitoring of water use in the catchment would greatly assist in better hydrological simulation, as such information can be used to develop an irrigation routine in the model.

Another issue of uncertainty is the configuration of the groundwater reservoirs. From the shape of the observed hydrograph it appears that the recession is not linear, but rather logarithmic or another non-linear function. The research using tracers (Camacho Suarez *et al.*, 2015) revealed that there are two distinct groundwater components in the Kaap outlet, which can be indicative of different reservoirs that operate with distinct dynamics. Shallow groundwater responds quickly in rainfall events, and has the highest contribution to flow, particularly when the antecedent moisture in the catchment is already high. The other groundwater component is from deeper sources, which could be the regional groundwater, recharged in the headwaters of the catchment. This component is responsible for sustaining the baseflow during most of the year. In the months of February to April, when the catchments are already quite wet, most of the runoff is generated through direct runoff (Chapter 5).

The HBV model (Bergström, 1992; Lindström *et al.*, 1997) was set up for the catchments using similar input data, but only vegetation and elevation band zones were used to discretize the model. Precipitation, temperature and potential evaporation were used to drive the model and automatic calibration was applied to get the best performing parameter sets. Results of this model (available in the Appendix A4) were comparable to the STREAM model outputs. The HBV was able to better simulate the water balance in the Kaap outlet, but the shape of the hydrographs and FDCs were better captured by the Stream model. This confirms that the distribution of input data, as well as the understanding of dominant runoff generation zones indeed assisted in informing model parameters and model simulation. However, both models still lack the complexity to fully capture the runoff processes occurring in the Kaap and tributaries.

The assessment of model performance should not only rely on statistical measures, but also on other aspects such as shape of the hydrograph, flow duration curves, among other hydrological signatures.

7.4.4 Limitations and gaps in process understanding

Even though great effort was made in selecting the best available input data, and making use of all available information and process understanding, there are some limitations and gaps in process understanding.

The representation of evaporation processes in the model is simplified, but is largely consistent with the water balance. The interception process is represented by a single daily threshold, and the interception storage is not simulated dynamically.

There is also the limitation of available data for potential evaporation. It is likely that the spatial variability of evaporation is higher than that simulated, given that stations used to interpolate climatic data were located in low altitudes. Furthermore, temporal variability of potential and actual evaporation, based on the vegetation cover growing stage and physiology occurs. These aspects were not captured in the current model configurations.

An attempt was made to use remote sensing actual evaporation products, such as ALEXI (Anderson *et al.*, 1997; Hain *et al.*, 2009), SSEBop (Senay *et al.*, 2013; Chen *et al.*, 2016) and CMERST (Guerschman *et al.*, 2009). However, these products have different time scales (weekly and monthly, respectively) and there were challenges with temporal interpolation of this data. We were able to downscale the ALEXI product and use it to drive the model, but the results were disappointing. Comparison at monthly and annual scale revealed that both ALEXI and CMERST products generally overestimate actual evaporation, whereas SSEBop results in underestimation. A potential way forward would be to use an ensemble product, and explore bias correction of the evaporation.

The runoff generation module of STREAM requires further development for model applications at daily time steps. The routing of runoff, the lateral flow process and the percolation were some of the gaps in the current model setup. In literature, most publications applying the STREAM model were done at monthly (Gerrits, 2005; Bouwer *et al.*, 2006; Winsemius *et al.*, 2006; Ward *et al.*, 2007) and weekly (Kiptala *et al.*, 2014) time steps. The reported daily model applications were also at much coarser spatial scale (Bouwer *et al.*, 2008; Ward *et al.*, 2008; Ward *et al.*, 2011).

7.5 Conclusions

This study combined hydrological modelling with mapping of dominant runoff generation processes, and a runoff signatures approach to improve the understanding of hydrological processes and runoff generation in a semi-arid African catchment.

Several data sources, parameter input values, and model structures were explored, in order to better understand the dominant processes in the catchment. Runoff response was sensitive to parameters related to the partitioning of rainfall between unsaturated and saturated zone *cr*, as well as the thresholds for initiation of quickflow *qc*. However, the inclusion of the feedback process from the saturated zone to the unsaturated zone, termed capillary rise, proved critical to improve model simulations. This was particularly the case for the Kaap and Queens catchments, which have a more diverse geology, coarser soils and hillslope zone. The Noordkaap and Suidkaap catchments have mostly fractured granite for bedrock, have deeper soils, and more plateaus, which results in more subsurface flow occurring.

The results of model simulations were analysed using the hydrological signatures framework as well as standard goodness of fit parameters. Annual runoff, seasonal runoff, flow duration curves and hydrographs of the different model runs were compared. The annual runoff showed that these catchments have high inter-annual variability, driven mostly by the variability of rainfall. The models were able to better simulate flows in wetter years (2010-2013) than in drier years (2004-2006). The seasonal flow analysis also revealed that there is strong seasonality in the flow generation. The capillary rise process in the model required a minimum threshold of initiation of the process (*GWSmin*) to avoid that the groundwater storage would run completely dry, which is not the case in the observed series of streamflow.

FDCs were the signature that best revealed the performance of different model simulations. In most cases the model was able to capture the slope of the FDC up to Q50/60, but missed the slope during the low flows. This was the case especially in the Kaap and Queens catchments. This finding reflects the importance of improving the representation of the evaporation and groundwater-surface water interaction processes, as well as water abstractions in the model setups, to better simulate the low flows.

Finally, the daily and monthly hydrographs were compared, and goodness of fit parameters computed between observed and modelled streamflow. Even though the goodness of fit results were average (0.75 to 0.84 Pearson R^2, 0.5 to 0.66 NSE in the best simulations), visual comparison shows that the models were able to capture the flow variability well, but missed the simulation of peak flows and overestimated baseflows.

A comparison was also made between the STREAM and HBV models. This yielded very similar results in terms of goodness of fit statistics for headwater catchments, but HBV performed better for the Kaap outlet. However, when the results were visually compared in terms of the various signatures used, the Stream model better captured the hydrograph shape, and the flow duration curve, particularly for baseflows.

This study clearly shows that there is no single model setup that can represent all the processes equally well for all the catchments. Due to the differences in landscape, geology, soils and land-use and land cover, different model configurations are better suited for each catchment. However, the distribution of input data, as well as the understanding of dominant runoff generation zones assisted in informing the Stream model. There is a benefit in combining process studies and modelling. The models highlight the shortcomings in process understanding, illustrating gaps in our knowledge. Process studies in this catchment assisted in filling some of this knowledge gaps, but other shortcomings were identified. Future improvements in the model should include the explicit accounting for irrigation and water transfers.

In terms of water management, the research findings reveal that the Queens and Kaap catchments are more sensitive to pollution, particularly during low flows, due to higher level of groundwater/surface water interactions. It is important to improve monitoring of water use, given the high impact of water abstractions in the catchment. The use of remote sensing products could assist in this, but more research is required for bias correction and calibration of products. Improvement in the calculation of actual evaporation is also required, as this constitutes the major component of the water balance, and there is high uncertainty in parameters used, and different evaporation products.

8

CONCLUSIONS AND RECOMMENDATIONS

8.1 General conclusions

The key objective of this PhD was to improve hydrological process understanding in the Incomati River basin, and the Kaap sub-catchment as a case study, using tracers and hydrological modelling, with the ultimate goal to better inform water management in the basin.

To achieve this aim, several tools were used.

- Hydroclimatic analysis of the catchment – detailed statistical analysis of rainfall and flow, using Indicators of Hydrological Alteration tool, to understand main drivers of hydroclimatic variability in the catchment;
- Tracer studies:
    - Hydrochemical snapshot sampling of the entire Incomati basin during two wet seasons and two dry seasons (2011 to 2013);
    - Intense event sampling in the Kaap catchment during 2013/2014 wet season;
    - Hydrograph separation at seasonal scale for the Kaap catchment with long term water quality data;
    - Hydrograph separation at event scale in the Kaap using multiple tracers;
- Hydrological modelling of the Kaap using STREAM model, with varying levels of complexity; and
- Analysis of multiple data sets of rainfall, evaporation, soil data, including remote sensing data.

This thesis was able to contribute to the improvement of understanding of hydrological processes, particularly runoff generation processes in a semi-arid river basin. Furthermore, the use of water quality data to bridge knowledge gaps in runoff generation processes was a significant contribution to the hydrological sciences. Several other gaps were highlighted, such as the disparity in available data (rainfall, streamflow, water quality, water use) in the Incomati River basin between riparian countries, the need for proper calibration and correction of remote sensing data before use for hydrological modelling, and the impacts of land use and land cover change on the water quantity and quality in the basin.

8.2 Main scientific contributions

Drivers of spatial and temporal variability of streamflow in the Incomati River Basin

A comprehensive analysis of variability and trends of streamflow, rainfall and land use changes was conducted for the semi-arid and transboundary Incomati River basin. The statistical analysis of rainfall data, which is a proxy for climate, revealed no consistent significant trend of increase or decrease for the studied period. The analysis of streamflow on the other end, using the tool Indicators of Hydrologic Alteration (IHA), revealed significant decreasing trends of streamflow indicators, particularly the median monthly flows of September and October, and low flow indicators. An analysis of land use changes was also conducted, using available secondary data, and were compared with mapped trends of streamflow and rainfall. Results showed that land use and flow regulation are the largest drivers of streamflow changes in the basin. Over the past 40 years, the areas under commercial forestry and irrigated agriculture have increased over four times, increasing the consumptive water use, particularly in the South African part of the basin.

This study shows that changes in flow regime in the Incomati basin are mostly driven by anthropogenic activities (e.g. irrigated agriculture, forestation, dam operation) and not by climate change. This means that great attention should be put into land use planning and management, and overall water management in the basin, to ensure sustainable use of water resources, whilst protecting the environment. This would also support mitigating climate change and coping with its impacts that are a recognised threat, particularly in the wider Southern African context.

Isotopic and hydrochemical river profile of the Incomati River Basin

The snapshot sampling in the Incomati basin for four seasons, as well as the analysis of long term secondary water quality data, revealed interesting patterns and trends.

Overall, electrical conductivity (EC) was identified as a good and significant indicator of status of hydrochemistry in the basin. EC data is more consistent in the database, and it is relatively cheap and easy to collect. Furthermore, EC is strongly correlated with major cations, and serves as a proxy of hydrochemistry.

An increasing trend of EC, major anions and cations, but a decreasing trend in water quality was observed from upstream headwaters to downstream in the Incomati Basin. The environmental isotopes also indicate more depleted water in headwaters (resembling more rain water) and more enriched water downstream, particularly in the Crocodile and Komati systems, which are more impacted by land use activities. The signal of irrigation return flows could be traced by elevated EC, high Sodium (Na^+) and enrichment of isotopes. Impact of reservoirs can be traced through enrichment of isotopes, due to the process of evaporation.

The Crocodile stream flow plays an important role in diluting polluted water flowing from the heavily polluted tributaries Elands and Kaap. The Sabie River likely plays a very important role of maintaining some basic water quality in the entire Incomati system.

Over time, increasing trends of several water quality parameters were also observed, and on many occasions, particularly in the Lower Komati and Crocodile catchments, the thresholds of water quality for industrial, irrigation and even domestic use were exceeded. The water quality parameters are strongly influenced by seasonality, and it is during the low flow season, and in drier years, that the deterioration of water quality in the basin is most critical.

This study shows once more that anthropogenic activities affect negatively water quality in the basin. While some stakeholders are already implementing measures to control water pollution, more emphasis is required to monitor and control pollution from various point and non-point sources. For this, it is critical that frequency of water quality monitoring is again increased, and some real time water quality sensors be installed at key/hot spot locations.

Hydrograph separation using tracers and digital filters to quantify runoff components in a semi-arid meso-scale catchment

Following the large scale water quality analysis of the Incomati Basin, a more in depth study was conducted in the Kaap catchment. This study tested a novel approach of using readily available secondary water quality data, particularly EC, to calibrate hydrograph separation using recursive digital filters, in order to quantify runoff components in the semi-arid catchment at daily, monthly and annual time scales. We also developed a method to optimize the baseflow index BFI_{max} parameter used in digital filter methods for hydrograph separation.

The chemical hydrograph separation indicates that the baseflow dominates the total flow, with contributions ranging from 50% in the wet season to 90% in the dry season. Hydrograph separation was then performed using Eckhardt's recursive digital filter, with daily streamflow data. The parameter BFI_{max} was calibrated for different sets of groundwater end members, using the chemical hydrograph separation for reference. The digital filter parameters are very sensitive, and their use without calibration is not recommended, as they can yield very different quantitative results. Optimal sets of α and BFI_{max} were identified for the studied catchments, which can serve as reference for further studies in the region or in semi-arid catchment elsewhere. In spite of the uncertainties in α and BFI_{max}, the digital filter hydrograph separation is very useful to interpolate and to extend baseflow estimates for periods without tracer data.

Another important finding of this study is the high contribution of baseflow to total flow during both wet and dry conditions. This means that the groundwater reservoirs respond quickly during storm events, which is important to consider for flood forecasting, environmental flow assessments, and for land use planning and management, in order to optimize/enhance groundwater recharge or prevent practices that compromise this.

The relevance of this analysis is that it allows for estimation of the baseflow component on a daily basis, from readily available streamflow and water quality data. Thus, these findings can be used to improve rainfall-runoff models, conjunctive groundwater management, river operations, and quantification of environmental flows where decisions regarding releases from dams and/or abstractions from rivers are done on a daily/weekly basis.

Understanding runoff processes in a semi-arid environment through isotope and hydrochemical hydrograph separations

A more intense study of runoff components was conducted in the semi-arid subtropical Kaap catchment, using environmental isotopes and hydrochemical tracers at event scale, for the 2013-2014 wet season. Hydrometric measurements and groundwater observations complemented the fieldwork. The suitability of isotope hydrograph separation was tested by comparing it to hydrochemical hydrograph separation showing no major differences between these tracers. Hydrograph separation showed that groundwater was the dominant runoff component for the wet season 2013-2014. Two-component hydrograph separation revealed groundwater contributions of between 64 and 98% of total runoff. An end-member mixing analysis (EMMA) suggested three runoff components, which are direct runoff, shallow and deep groundwater components. Direct runoff, defined as the direct precipitation on the stream, channel and overland flow, contributed up to 41% of total runoff during wet catchment conditions. Shallow groundwater defined as the soil water and near surface water component contributed up to 45% of total runoff, and deep groundwater contributed up to 84% of total runoff.

A strong correlation between direct runoff generation and antecedent precipitation conditions was found for the four studied events. These findings suggest that direct runoff is enhanced by wetter conditions in the catchment that trigger saturation excess overland flow as observed in the hydrograph separations. The understanding of runoff generation mechanisms in the Kaap catchment contributes to the limited number of hydrological processes studies and in particular to hydrograph separation studies in semi-arid and tropical regions, which are generally poorly understood and characterized. The study demonstrated the potential to improve water management in semi-arid watersheds, where flow is often limited or low, by revealing were runoff is generated, what are hot spots for surface and groundwater interactions, and where more protection of water resources is required.

Improved process representation in the simulation of the hydrology of a meso-scale semi-arid catchment

To consolidate the understanding of hydrological processes in the Kaap catchment, novel mapping of dominant runoff generation zones was combined with hydrological modelling of the catchment. Several new data sources, parameter input values, and model structures were explored, in order to better understand the dominant processes in the catchment. Dominant runoff processes were mapped using a simplified Height Above the Nearest Drainage (HAND) approach combined with geology, which is a novel approach in the region. The Prediction in Ungauged Basins (PUB) framework of runoff signatures was used to analyse the results of applying the (open source) STREAM model.

Results show that in the headwater sub-catchments of Noordkaap and Suidkaap, plateaus dominate, associated with slow flow processes, with high infiltration, percolation and groundwater recharge and less quickflow. Therefore, these catchments have high baseflow components, and are likely the main recharge zone for regional groundwater in the Kaap. In the Queens sub-catchment, in contrast, hillslopes associated with intermediate and fast flow processes dominate. However, this catchment still has a relatively strong baseflow component, but it seems to be more impacted by evaporation depletion, due to different soils and geology, especially in drier years. At the Kaap outlet, the model indicates that hillslopes are important for runoff generation in the catchment; intermediate and fast flow processes dominate and most runoff is generated through direct runoff and shallow groundwater, particularly in wetter months and years. There is a high impact of water abstractions and evaporation during the dry season, affecting low flows. Results also indicate that the root zone storage and the parameters of effective rainfall separation (between unsaturated and saturated zone), quickflow coefficient and capillary rise, were very sensitive in the model. The inclusion of capillary rise (feedback from saturated to unsaturated zone) greatly improved simulation results.

An attempt was made to use remote sensing actual evaporation products, such as ALEXI, SSEBop and CMERST. However, these products have different time scales (weekly and monthly, respectively) and there were challenges with temporal interpolation of this data. It was possible to downscale the ALEXI product and use it to drive the model, but the results were disappointing. Comparison at monthly and annual scale revealed that both ALEXI and CMERST products generally overestimate actual evaporation, whereas SSEBop underestimates it. A potential way forward would be to use an ensemble product, and explore bias correction of the evaporation.

A comparison was also made between the STREAM and HBV models. This yielded very similar results in terms of goodness of fit statistics for headwater catchments, but HBV performed better for the Kaap outlet. However, when the results were visually compared, the STREAM model better captured the hydrograph shape and the flow duration curve, particularly during baseflow.

This study clearly shows that there is no single model setup that can represent all the processes equally well for all the catchments. Due to the differences in landscape, geology, soils and land-use and land cover, different model configurations are better suited for each catchment. However, the distribution of input data, as well as the understanding of dominant runoff generation zones assisted in informing the STREAM model (definition of parameters related with quickflow coefficient, rainfall partition, maximum storage in unsaturated and groundwater zones, and capillary rise threshold). There is a benefit in combining process studies and modelling. The models highlight the shortcomings in process understanding, illustrating gaps in our knowledge. Process studies in this catchment assisted in filling some of these knowledge gaps, while new shortcomings were identified.

8.3 Novelty of this PhD

This PhD research sheds light on the understanding of hydrological processes, particularly runoff generation processes in a semi-arid sub-tropical catchment in Southern Africa. Novel methods are applied to this region, to understand the hydroclimatic variability, the drivers of river flow regime alteration, the dominant runoff generation processes as well as new data sources for hydrological modelling.

A new method to use hydrochemistry data in order to calibrate recursive digital filters for hydrograph separation was tested, and runoff components were quantified for the semi-arid subtropical Kaap catchment in South Africa at monthly and annual scale. This approach was novel in the region, and provides a road map for further exploration of water quality data to improve understanding of hydrological processes in the region. The results of the hydrograph separation can be used for operational management of environmental flows, and to further inform hydrological models in the region.

Furthermore, new data sets were generated from water quality data and environmental isotopes for the Incomati basin, and the Kaap catchment in particular. These datasets enable description of water quality patterns in the Incomati and improved the understanding of runoff generation processes in the Kaap catchment. Four events were intensely sampled in the Kaap catchment during a wet season, which enabled understanding of the flow generation dynamics in this catchment. The role of antecedent precipitation for runoff generation was highlighted, as well as the prevalence of groundwater in runoff generation. Gaps in understanding of groundwater components and sources were also identified.

The application of the STREAM model for the Kaap catchment was also innovative. Few applications of the model at daily time scale are reported in literature. This study provided the parameterization of the model for a semi-arid catchment, as well as a method to better inform model parameters based on novel landscape classification. Furthermore, new datasets of remote sensing data to drive the model were explored. These datasets are very relevant for the case of transboundary basins, such as the Incomati, where discrepancy in data availability between different countries can hamper the modelling of water resources of the whole basin.

Furthermore, the findings of this thesis provide a basis for better management of water resources in the region, which has been assured through stakeholder involvement (e.g. the reference group meetings, various RISKOMAN project meetings, cooperation with catchment management agency, joint fieldwork, etc).

8.4 Recommendations for future research

This study evaluated several data sources for hydrometric, water quality and physiographic characteristics of the Incomati Basin. A challenge identified is the disparity in data availability between the different riparian countries. For example, in Mozambique only two flow gauges had reliable data for the streamflow analysis. The rivers Massintonto, Uanetse and Mazimchopes in Mozambique do not have active flow gauges. There is a need to strengthen the monitoring network in the basin, particularly for flow, rainfall and groundwater.

There is also an alarming trend of a reduction in the number of active rainfall stations and a reduction in the water quality sampling frequency in the Incomati overall. It is critical to evaluate the value of data, and establish a network with appropriate spatial/temporal coverage. It is also important to have a common database for the basin, and similar protocols for data collection and reporting.

From the analysis of drivers of streamflow variability, it was clear that the Sabie catchment had different patterns compared to Komati and Crocodile. It had less negative trends in hydrological indicators, and it was more compliant with environmental flow requirements. It is likely that the strategic adaptive management approach adopted by the Kruger National Park and Inkomati-Usuthu Catchment Management Agency, are assisting in maintaining environmental flows in the Sabie, and could be further employed in the basin, as a management framework. An evaluation of the application of strategic adaptive management at the Incomati basin scale should be further investigated.

Considering the high spatial variability in the observed streamflow changes, specific tailor-made interventions are needed for the most affected sub-catchments and main catchments. Future investigations should conduct a careful basin-wide assessment of benefits derived from water use including the consideration of environmental requirements, and assess the first priority water uses, including commercial forest plantations; the latter are, *de facto*, not *de jure*, priority users, as they take the water first.

The process studies using tracers were limited by the water quality data available. There was no detailed data on several water sources, such as return flows from irrigation, industries and settlements. It is important to further investigate the influence of such sources in the signature of runoff end member components. Further research is necessary to make as much as possible a clear distinction between surface runoff and shallow groundwater components. In addition, process studies focusing on the dry season should quantify the dependency of runoff generation on soil moisture and vegetation.

Recursive digital filters are very useful and inexpensive to quantify baseflow contributions. However, given the high sensitivity of digital filter parameters, it is important that the calibration of parameters is done prior to operational use of baseflow estimates. Furthermore, more validation studies should be conducted in other semi-arid catchments to assess if regionalization (transfer in space) of recursive digital filter parameters is possible, ideally using high frequency water quality data. This could be achieved with the installation of real time EC sensors in selected catchments to enable detailed calibration.

The rainfall input greatly impacts the runoff generation. The regionalization of rainfall based on available stations using the IDW method in the Kaap could have induced a greater number of rainy days and reduced the magnitude of rain events considering that the rainfall is mainly originating from convective rain cells. This results in basin-wide higher interception rates, and underestimation of flow peaks. It would be useful to conduct a comprehensive study on bias correction, downscaling and calibration of remote sensing precipitation data to use as input for hydrological modelling in the Kaap catchment and also in the Incomati basin as a whole. This would be very valuable at the basin scale, given the uneven spatial coverage of the monitoring network.

The representation of evaporation processes in the STREAM model used is simplified, but is largely consistent with the annual water balance. The interception process is represented by a single daily threshold, while the interception storage is not simulated dynamically. Although temporal variability of potential and actual evaporation, based on the vegetation cover growing stage and physiology, does occur, this could not be captured by the current model configurations, but should be investigated in future research. Furthermore, the use of remote sensing actual evaporation products should be explored, either by testing ensemble products to drive the model, or by bias correction of the evaporation products.

Future improvements in the model should include the explicit accounting for irrigation and water transfers. The runoff generation module of STREAM requires further development for model applications at daily time steps. The routing of runoff, the lateral flow process and the percolation were some of the shortcomings in the current model setup.

Finally, it will be relevant to investigate how the better process understanding with spatially distributed consideration of runoff generation area can assist in better land use planning, water allocation and water management in the basin. For instance, through scenario development, assist policy and decision makers (including stakeholders from the different countries) make the strategic and operational plans.

9 REFERENCES

Abiye T, Verhagen B, Freese C, Harris C, Orchard C, Van Wyk E, Tredoux G, Pickles J, Kollongei J, Xiao L. 2013. The use of isotope hydrology to characterize and assess water resources in South (ern) Africa. WRC Report No. TT570/13, Pretoria. 211pp.

Acocks JPH. 1988. Veld Types of South Africa, Memoirs of the Botanical Survey of South Africa. Botanical Research Institute, Dept. of Agriculture and Water Supply (South Africa), 57, 146 pp.

Aerts JCJH, Kriek M, Schepel M. 1999. STREAM (Spatial tools for river basins and environment and analysis of management options): 'set up and requirements'. Physics and Chemistry of the Earth, Part B: Hydrology, Oceans and Atmosphere, **24**: 591-595. DOI: http://dx.doi.org/10.1016/S1464-1909(99)00049-0.

Allen RG, Pereira LS, Raes D, Smith M. 1998. Crop evapotranspiration-Guidelines for computing crop water requirements-FAO Irrigation and drainage paper 56. FAO, Rome, **300**: 6541.

Anderson MC, Norman JM, Diak GR, Kustas WP, Mecikalski JR. 1997. A two-source time-integrated model for estimating surface fluxes using thermal infrared remote sensing. Remote Sensing of Environment, **60**: 195-216. DOI: https://doi.org/10.1016/S0034-4257(96)00215-5.

Aurecon. 2010. Tripartite Technical Committee (TPTC) between Moçambique, South Africa and Swaziland. IAAP 10: Consulting Services for System Operating Rules for the Incomati and Maputo Watercourses. System Operating Rules: Status Report.

Bailey A, Pitman W. 2015. Water Resources of South Africa, 2012 Study : User Guide. Water Research Comission.

Bakhit MSI. 2014. Assessing the feasibility of the conjunctive use of surface water and groundwater during drought in the context of sustainable development - The case of the Lomati Catchment. In: MSc Thesis WM-WRM. 14.11, UNESCO-IHE, pp: 134.

Bazemore DE, Eshleman KN, Hollenbeck KJ. 1994. The Role of Soil-Water in Stormflow Generation in a Forested Headwater Catchment - Synthesis of Natural Tracer and Hydrometric Evidence. Journal of Hydrology, **162**: 47-75. DOI: Doi 10.1016/0022-1694(94)90004-3.

Bergström S. 1992. The HBV model: Its structure and applications. Swedish Meteorological and Hydrological Institute.

Beumer J, Mallory S. 2014. Water Requirements and Availability Reconciliation Strategy for the Mbombela Municipal Area. Final Reconcialiation Strategy. Department of Water Affairs.

Beven KJ. 2012. Rainfall-runoff modelling: the primer. John Wiley & Sons.

Birkel C, Soulsby C, Ali G, Tetzlaff D. 2014a. Assessing the cumulative impacts of hydropower regulation on the flow characteristics of a large atlantic Salmon river system. River Research and Applications, **30**: 456-475. DOI: 10.1002/rra.2656.

Birkel C, Soulsby C, Tetzlaff D. 2014b. Developing a consistent process-based conceptualization of catchment functioning using measurements of internal state variables. Water Resources Research, **50**: 3481-3501. DOI: 10.1002/2013wr014925.

Birkel C, Tetzlaff D, Dunn SM, Soulsby C. 2010. Towards a simple dynamic process conceptualization in rainfall–runoff models using multi-criteria calibration and tracers in temperate, upland catchments. Hydrological Processes, **24**: 260-275. DOI: 10.1002/hyp.7478.

Blöschl G, Sivapalan M, Wagener T, Viglione A, Savenije H. 2013a. Runoff prediction in ungauged basins. Cambridge University Press.

Blöschl G, Sivapalan M, Wagener T, Viglione A, Savenije H. 2013b. Runoff Prediction in Ungauged Basins: Synthesis Across Processes, Places and Scales. Cambridge University Press.

Bouwer LM, Aerts JCJH, Droogers P, Dolman AJ. 2006. Detecting the long-term impacts from climate variability and increasing water consumption on runoff in the Krishna river basin (India). Hydrol. Earth Syst. Sci., **10**: 703-713. DOI: 10.5194/hess-10-703-2006.

Bouwer LM, Biggs TW, Aerts JCJH. 2008. Estimates of spatial variation in evaporation using satellite-derived surface temperature and a water balance model. Hydrological Processes, **22**: 670-682. DOI: doi:10.1002/hyp.6636.

Bunn SE, Arthington AH. 2002. Basic Principles and Ecological Consequences of Altered Flow Regimes for Aquatic Biodiversity. Environmental Management, **30**: 492-507. DOI: 10.1007/s00267-002-2737-0.

Burns DA. 2002. Stormflow-hydrograph separation based on isotopes: the thrill is gone - what's next? Hydrological Processes, **16**: 1515-1517. DOI: Doi 10.1002/Hyp.5008.

Burns DA, McDonnell JJ, Hooper RP, Peters NE, Freer JE, Kendall C, Beven K. 2001. Quantifying contributions to storm runoff through end-member mixing analysis and hydrologic measurements at the Panola Mountain Research Watershed (Georgia, USA). Hydrological Processes, **15**: 1903-1924. DOI: Doi 10.1002/Hyp.246.

Butterworth JA, Schulze RE, Simmonds LP, Moriarty P, Mugabe F. 1999. Hydrological processes and water resources management in a dryland environment IV: Long-term groundwater level fluctuations due to variation in rainfall. Hydrol. Earth Syst. Sci. HESS, **3**. DOI: 0.5194/hess-3-353-1999.

Buttle JM. 1994. Isotope hydrograph separations and rapid delivery of pre-event water from drainage basins. Progress in Physical Geography, **18**: 16-41. DOI: 10.1177/030913339401800102.

Camacho Suarez VV, Saraiva Okello AML, Wenninger JW, Uhlenbrook S. 2015. Understanding runoff processes in a semi-arid environment through isotope and hydrochemical hydrograph separations. Hydrol. Earth Syst. Sci., **19**: 4183-4199. DOI: 10.5194/hess-19-4183-2015.

Camarasa-Belmonte AM, Soriano J. 2014. Empirical study of extreme rainfall intensity in a semi-arid environment at different time scales. Journal of Arid Environments: 63-71.

Capell R, Tetzlaff D, Malcolm IA, Hartley AJ, Soulsby C. 2011. Using hydrochemical tracers to conceptualise hydrological function in a larger scale catchment draining contrasting geologic provinces. Journal of Hydrology, **408**: 164-177. DOI: http://dx.doi.org/10.1016/j.jhydrol.2011.07.034.

Capell R, Tetzlaff D, Soulsby C. 2012. Can time domain and source area tracers reduce uncertainty in rainfall-runoff models in larger heterogeneous catchments? Water Resources Research, **48**: W09544. DOI: 10.1029/2011wr011543.

Carmo Vaz Á, Lopes Pereira A. 2000. The Incomati and Limpopo international river basins: a view from downstream. Water Policy, **2**: 99-112.

Carmo Vaz A, Van Der Zaag P. 2003. Sharing the Incomati waters: cooperation and competition in the balance. In: Water Policy, pp: 349-368.

Cartwright I, Gilfedder B, Hofmann H. 2014. Contrasts between estimates of baseflow help discern multiple sources of water contributing to rivers. Hydrol. Earth Syst. Sci., **18**: 15-30. DOI: 10.5194/hess-18-15-2014.

CAWMA. 2007. Water for food, water for life: A comprehensive assessment of water management in agriculture. Summary., London: Earthscan, and Colombo: International Water Management Institute.

Chen M, Senay GB, Singh RK, Verdin JP. 2016. Uncertainty analysis of the Operational Simplified Surface Energy Balance (SSEBop) model at multiple flux tower sites. Journal of Hydrology, **536**: 384-399. DOI: https://doi.org/10.1016/j.jhydrol.2016.02.026.

Christophersen N, Hooper RP. 1992. Multivariate analysis of stream water chemical data: The use of principal components analysis for the end-member mixing problem. Water Resources Research, **28**: 99-107. DOI: 10.1029/91wr02518.

Davies EGR, Simonovic SP. 2011. Global water resources modeling with an integrated model of the social-economic-environmental system. Advances in Water Resources, **34**: 684-700.

De Lange WJ, Wise RM, Forsyth GG, Nahman A. 2010. Integrating socio-economic and biophysical data to support water allocations within river basins: An example from the Inkomati Water Management Area in South Africa. Environmental Modelling & Software, **25**: 43-50.

De Winnaar G, Jewitt G. 2010. Ecohydrological implications of runoff harvesting in the headwaters of the Thukela River basin, South Africa. Physics and Chemistry of the Earth, Parts A/B/C, **35**: 634-642. DOI: http://dx.doi.org/10.1016/j.pce.2010.07.009.

de Wit MJ, Furnes H, Robins B. 2011. Geology and tectonostratigraphy of the Onverwacht Suite, Barberton Greenstone Belt, South Africa. Precambrian Research, **186**: 1-27. DOI: http://dx.doi.org/10.1016/j.precamres.2010.12.007.

Deksissa T, Ashton PJ, Vanrolleghem PA. 2003. Control options for river water quality improvement: A case study of TDS and inorganic nitrogen in the Crocodile River (South Africa). Water SA, **29**: 209-218.

Diamond RE, Jack S. 2018. Evaporation and abstraction determined from stable isotopes during normal flow on the Gariep River, South Africa. Journal of Hydrology, **559**: 569-584. DOI: https://doi.org/10.1016/j.jhydrol.2018.02.059.

Dlamini EM. 2007. Decision support systems for managing the water resources of the Komati River Basin. International Journal of River Basin Management, **5**: 179 - 187.

DWAF. 1996a. Department of Water Affairs and Forestry, South African Water Quality Guidelines (second edition). Volume 1: Domestic Use.

DWAF. 1996b. Department of Water Affairs and Forestry. South African Water Quality Guidelines (second edition), Volume 3: Industrial Use.

DWAF. 1996c. Department of Water Affairs and Forestry. South African Water Quality Guidelines (second edition). Volume 4: Agricultural Use: Irrigation.

DWAF. 2003a. Department of Water Affairs and Forestry, South Africa. Inkomati Water Management Area: Overview of Water Resources Availability and Utilisation.

DWAF. 2003b. Department of Water Affairs and Forestry, South Africa. Sabie River Catchment: Operating Rules and Decision Support Models for Management of the Surface Water Resources.

DWAF. 2004. Department of Water Affairs and Forestry, South Africa. Internal Strategic Perspectives: Inkomati Water Management Area – Version 1 (March 2004). Tlou & Matji (Pty) Ltd.

DWAF. 2009a. The Development of a Real-Time Decision Support System (DSS) for the Crocodile East River System. Main Report.

DWAF. 2009b. Inkomati Water Availability Assessment Study. Hydrology of Crocodile River Volume 1.

DWAF. 2009c. Inkomati Water Availability Assessment Study. Water Requirements Volume 1.

DWAF. 2009d. Inkomati Water Availability Assessment Study: Hydrology of Sabie River Volume 1.

DWAF. 2009e. Inkomati Water Availability Assessment Study: Main Report.

DWAF. 2009f. Inkomati Water Availability Assessment Study: Water Quality Situation.

Dye PJ. 1996. Response of Eucalyptus grandis trees to soil water deficits. Tree Physiology, **16**: 233-238. DOI: 10.1093/treephys/16.1-2.233.

Eckhardt K. 2005. How to construct recursive digital filters for baseflow separation. Hydrological Processes, **19**: 507-515. DOI: 10.1002/hyp.5675.

Eckhardt K. 2008. A comparison of baseflow indices, which were calculated with seven different baseflow separation methods. Journal of Hydrology, **352**: 168-173. DOI: http://dx.doi.org/10.1016/j.jhydrol.2008.01.005.

Eckhardt K. 2012. Technical Note: Analytical sensitivity analysis of a two parameter recursive digital baseflow separation filter. Hydrol. Earth Syst. Sci., **16**: 451-455. DOI: 10.5194/hess-16-451-2012.

Falkenmark M. 1997. Meeting water requirements of an expanding world population. Philosophical Transactions of the Royal Society of London. Series B: Biological Sciences, **352**: 929-936. DOI: 10.1098/rstb.1997.0072.

Falkenmark M, Molden D. 2008. Wake up to realities of river basin closure. Int. J. Water Resour. Dev. International Journal of Water Resources Development, **24**: 201-215.

Fanta B, Zaake BT, Kachroo RK. 2001. A study of variability of annual river flow of the southern African region. Hydrological Sciences Journal, **46**: 513-524. DOI: 10.1080/02626660109492847.

Farmer D, Sivapalan M, Jothityangkoon C. 2003. Climate, soil, and vegetation controls upon the variability of water balance in temperate and semiarid landscapes: Downward approach to water balance analysis. Water Resources Research, **39**: 1035.

Frisbee MD, Phillips FM, White AF, Campbell AR, Liu F. 2013. Effect of source integration on the geochemical fluxes from springs. Applied Geochemistry, **28**: 32-54. DOI: http://dx.doi.org/10.1016/j.apgeochem.2012.08.028.

Funk C, Peterson P, Landsfeld M, Pedreros D, Verdin J, Shukla S, Husak G, Rowland J, Harrison L, Hoell A, Michaelsen J. 2015. The climate hazards infrared precipitation with stations—a new environmental record for monitoring extremes. Scientific Data, **2**: 150066. DOI: 10.1038/sdata.2015.66.

Gallego-Ayala J, Juízo D. 2012. Performance evaluation of River Basin Organizations to implement integrated water resources management using composite indexes. Physics and Chemistry of the Earth, Parts A/B/C, **50–52**: 205-216. DOI: http://dx.doi.org/10.1016/j.pce.2012.08.008.

Gallego-Ayala J, Juízo D. 2014. Integrating Stakeholders' Preferences into Water Resources Management Planning in the Incomati River Basin. Water Resources Management, **28**: 527-540. DOI: 10.1007/s11269-013-0500-3.

Gao H, Hrachowitz M, Fenicia F, Gharari S, Savenije HHG. 2014. Testing the realism of a topography-driven model (FLEX-Topo) in the nested catchments of the Upper Heihe, China. Hydrol. Earth Syst. Sci., **18**: 1895-1915. DOI: 10.5194/hess-18-1895-2014.

Genereux D. 1998. Quantifying uncertainty in tracer-based hydrograph separations. Water Resources Research, **34**: 915-919. DOI: Doi 10.1029/98wr00010.

Gerrits A. 2005. Hydrological modelling of the Zambezi catchment for gravity measurements. In: Water Resources, University of Technology.

Gharari S, Hrachowitz M, Fenicia F, Savenije HHG. 2011. Hydrological landscape classification: investigating the performance of HAND based landscape classifications in a central European meso-scale catchment. Hydrol. Earth Syst. Sci., **15**: 3275-3291. DOI: 10.5194/hess-15-3275-2011.

Gonzales AL, Nonner J, Heijkers J, Uhlenbrook S. 2009. Comparison of different base flow separation methods in a lowland catchment. Hydrol. Earth Syst. Sci., **13**: 2055-2068. DOI: 10.5194/hess-13-2055-2009.

Gröning M, Lutz HO, Roller-Lutz Z, Kralik M, Gourcy L, Pöltenstein L. 2012. A simple rain collector preventing water re-evaporation dedicated for 18-O and Deuterium analysis of cumulative precipitation samples. Journal of Hydrology: 195–200.

Group SCW, Macvicar C. 1991. Soil classification: A taxonomic system for South Africa. Department of Agricultural Development.

Guerschman JP, Van Dijk AIJM, Mattersdorf G, Beringer J, Hutley LB, Leuning R, Pipunic RC, Sherman BS. 2009. Scaling of potential evapotranspiration with MODIS data reproduces flux observations and catchment water balance observations across Australia. Journal of Hydrology, **369**: 107-119. DOI: https://doi.org/10.1016/j.jhydrol.2009.02.013.

Gupta HV, Kling H, Yilmaz KK, Martinez GF. 2009. Decomposition of the mean squared error and NSE performance criteria: Implications for improving hydrological modelling. Journal of Hydrology, **377**: 80-91. DOI: http://dx.doi.org/10.1016/j.jhydrol.2009.08.003.

Guzman J, Chu M. 2004. SPELL-stat v 15110 B. Grupo en Prediccion y Modelamiento Hidroclimatico Universidad Industrial de Santander CHMaPG, Industrial University of Santander, Colombia (ed.).

Hain CR, Mecikalski JR, Anderson MC. 2009. Retrieval of an available water-based soil moisture proxy from thermal infrared remote sensing. Part I: Methodology and validation. Journal of Hydrometeorology, **10**: 665-683. DOI: 10.1175/2008jhm1024.1.

Hall FR. 1968. Base-Flow Recessions—A Review. Water Resources Research, **4**: 973-983. DOI: 10.1029/WR004i005p00973.

Hellegers PJGJ, Perry CJ, Bastiaanssen WGM. 2009. Combining remote sensing and economic analysis to support decisions that affect water productivity. Irrigation Science, **27**: 243-251.

Hellegers PJGJ, Soppe R, Perry CJ, Bastiaanssen WGM. 2010. Remote Sensing and Economic Indicators for Supporting Water Resources Management Decisions. Water Resources Management, **24**: 2419-2436.

Hengl T, de Jesus JM, MacMillan RA, Batjes NH, Heuvelink GB, Ribeiro E, Samuel-Rosa A, Kempen B, Leenaars JG, Walsh MG. 2014. SoilGrids1km—global soil information based on automated mapping. PLoS One, **9**: e105992.

Hengl T, Heuvelink GB, Kempen B, Leenaars JG, Walsh MG, Shepherd KD, Sila A, MacMillan RA, de Jesus JM, Tamene L. 2015. Mapping soil properties of Africa at 250 m resolution: Random forests significantly improve current predictions. PLoS ONE, **10**: e0125814.

Hengl T, Mendes de Jesus J, Heuvelink GBM, Ruiperez Gonzalez M, Kilibarda M, Blagotić A, Shangguan W, Wright MN, Geng X, Bauer-Marschallinger B, Guevara MA, Vargas R, MacMillan RA, Batjes NH, Leenaars JGB, Ribeiro E, Wheeler I, Mantel S, Kempen B. 2017. SoilGrids250m: Global gridded soil information based on machine learning. PLoS ONE, **12**: e0169748. DOI: 10.1371/journal.pone.0169748.

Hessler AM, Lowe DR. 2006. Weathering and sediment generation in the Archean: An integrated study of the evolution of siliciclastic sedimentary rocks of the 3.2 Ga Moodies Group, Barberton Greenstone Belt, South Africa. Precambrian Research, **151**: 185-210. DOI: http://dx.doi.org/10.1016/j.precamres.2006.08.008.

Hoguane AM, Antonio MHP. 2016. The Hydrodynamics of the Incomati Estuary – An Alternative Approach to Estimate the Minimum Environmental Flow. In: Estuaries: A Lifeline of Ecosystem Services in the Western Indian Ocean, Diop S, Scheren P, Ferdinand Machiwa J (eds.) Springer International Publishing, pp: 289-300.

Hrachowitz M, Bohte R, Mul ML, Bogaard TA, Savenije HHG, Uhlenbrook S. 2011. On the value of combined event runoff and tracer analysis to improve understanding of catchment functioning in a data-scarce semi-arid area. Hydrol. Earth Syst. Sci., **15**: 2007-2024. DOI: 10.5194/hess-15-2007-2011.

Hrachowitz M, Savenije HHG, Blöschl G, McDonnell JJ, Sivapalan M, Pomeroy JW, Arheimer B, Blume T, Clark MP, Ehret U, Fenicia F, Freer JE, Gelfan A, Gupta HV, Hughes DA, Hut RW, Montanari A, Pande S, Tetzlaff D, Troch PA, Uhlenbrook S, Wagener T, Winsemius HC, Woods RA, Zehe E, Cudennec C. 2013. A decade of Predictions in Ungauged Basins (PUB)—a review. Hydrological Sciences Journal: 1-58. DOI: 10.1080/02626667.2013.803183.

Hrachowitz M, Soulsby C, Tetzlaff D, Dawson JJC, Dunn SM, Malcolm IA. 2009. Using long-term data sets to understand transit times in contrasting headwater catchments. Journal of Hydrology, **367**: 237-248. DOI: DOI 10.1016/j.jhydrol.2009.01.001.

Hu Y, Maskey S, Uhlenbrook S, Zhao H. 2011. Streamflow trends and climate linkages in the source region of the Yellow River, China. Hydrological Processes, **25**: 3399-3411. DOI: 10.1002/hyp.8069.

Hughes D. 2007. Modelling semi-arid and arid hydrology and water resources - the southern African experience. Chapter 3. In: Hydrological Modelling in Arid and Semi-Arid Areas, Cambridge University Press.

Hughes DA. 2010. Unsaturated zone fracture flow contributions to stream flow: evidence for the process in South Africa and its importance. Hydrological Processes, **24**: 767-774. DOI: 10.1002/hyp.7521.

Hughes DA. 2016. Hydrological modelling, process understanding and uncertainty in a southern African context: lessons from the northern hemisphere. Hydrological Processes, **30**: 2419-2431. DOI: 10.1002/hyp.10721.

Hughes DA, Hannart P, Watkins D. 2003. Continuous baseflow separation from time series of daily and monthly streamflow data. Water SA, **29**: 43-48.

Hughes DA, Jewitt G, Mahé G, Mazvimavi D, Stisen S. 2015. A review of aspects of hydrological sciences research in Africa over the past decade. Hydrological Sciences Journal: 1-15. DOI: 10.1080/02626667.2015.1072276.

Hughes DA, Mallory SJL. 2008. Including environmental flow requirements as part of real-time water resource management. River Research and Applications, **24**: 852-861. DOI: 10.1002/rra.1101.

Hughes DA, Tshimanga RM, Tirivarombo S, Tanner J. 2014. Simulating wetland impacts on stream flow in southern Africa using a monthly hydrological model. Hydrological Processes, **28**: 1775-1786. DOI: 10.1002/hyp.9725.

Hughes JD, Khan S, Crosbie RS, Helliwell S, Michalk DL. 2007. Runoff and solute mobilization processes in a semiarid headwater catchment. Water Resources Research, **43**.

Hümann M, Müller C. 2013. Improving the GIS-DRP approach by means of delineating runoff characteristics with new discharge relevant parameters. ISPRS International Journal of Geo-Information, **2**: 27.

ICMA. 2010. The Inkomati Catchment Management Strategy: A First Generations Catchment Management Strategy for the Inkomati Water Management Area., Inkomati Catchment Management Agency.

Jackson B. 2014. An adaptive operational water resources management framework for the Crocodile river catchment, South Africa. MSc Thesis. In: Centre for Water Resources Research, University of KwaZulu-Natal, South Africa.

Jarmain C, Dost RJJ, De Bruijn E, Ferreira F, Schaap O, Bastiaanssen WGM, Bastiaanssen F, van Haren I, van Haren IJ, Wayers T, Ribeiro D, Pelgrum H, Obando E, Van Eekelen MW. 2013. Spatial Hydro-meteorological data for transparent and equitable water resources management in the Incomati catchment. Report to the Water Research Commission., WRC (ed.).

Jewitt G. 2006a. Integrating blue and green water flows for water resources management and planning. Physics and Chemistry of the Earth, Parts A/B/C, **31**: 753-762. DOI: http://dx.doi.org/10.1016/j.pce.2006.08.033.

Jewitt G. 2006b. Water and Forests. In: Encyclopedia of Hydrological Sciences, John Wiley & Sons, Ltd.

Jewitt GPW. 2002. The 8%-4% debate: Commercial afforestation and water use in South Africa. Southern African Forestry Journal, **194**: 1-5. DOI: 10.1080/20702620.2002.10434586.

Jewitt GPW, Garratt JA, Calder IR, Fuller L. 2004. Water resources planning and modelling tools for the assessment of land use change in the Luvuvhu Catchment, South Africa. Physics and Chemistry of the Earth, Parts A/B/C, **29**: 1233-1241.

Jewitt GPW, Görgens AHM. 2000. Scale and model interfaces in the context of integrated water resources management for the rivers of the Kruger National Park., Water Research Commission.

JIBS. 2001. Joint Incomati Basin Study Report. Phase 2. Consultec Report No: C14-99MRF/BKS ACRES.

Keller J, Keller A, Davids G. 1998. River basin development phases and implications of closure. Journal of Applied Irrigation Science, **Vol. 33**: pp. 145-163.

Kennedy CD, Bataille C, Liu Z, Ale S, VanDeVelde J, Roswell CR, Bowling LC, Bowen GJ. 2012. Dynamics of nitrate and chloride during storm events in agricultural catchments with different subsurface drainage intensity (Indiana, USA). Journal of Hydrology, **466-467**: 1-10. DOI: 10.1016/j.jhydrol.2012.05.002.

Kiptala JK, Mul ML, Mohamed YA, van der Zaag P. 2014. Modelling stream flow and quantifying blue water using a modified STREAM model for a heterogeneous, highly utilized and data-scarce river basin in Africa. Hydrol. Earth Syst. Sci., **18**: 2287-2303. DOI: 10.5194/hess-18-2287-2014.

Klaus J, McDonnell JJ. 2013. Hydrograph separation using stable isotopes: Review and evaluation. Journal of Hydrology, **505**: 47-64. DOI: http://dx.doi.org/10.1016/j.jhydrol.2013.09.006.

Kling H, Fuchs M, Paulin M. 2012. Runoff conditions in the upper Danube basin under an ensemble of climate change scenarios. Journal of Hydrology, **424–425**: 264-277. DOI: http://dx.doi.org/10.1016/j.jhydrol.2012.01.011.

Kronholm SC, Capel PD. 2015. A comparison of high-resolution specific conductance-based end-member mixing analysis and a graphical method for baseflow separation of four streams in hydrologically challenging agricultural watersheds. Hydrological Processes, **29**: 2521-2533. DOI: 10.1002/hyp.10378.

Kruger AC, Shongwe S. 2004. Temperature trends in South Africa: 1960–2003. International Journal of Climatology, **24**: 1929-1945. DOI: 10.1002/joc.1096.

Laudon H, Slaymaker O. 1997. Hydrograph separation using stable isotopes, silica and electrical conductivity: an alpine example. Journal of Hydrology, **201**: 82-101. DOI: http://dx.doi.org/10.1016/S0022-1694(97)00030-9.

LeMarie M, van der Zaag P, Menting G, Baquete E, Schotanus D. 2006. The use of remote sensing for monitoring environmental indicators: The case of the Incomati estuary, Mozambique. Physics and Chemistry of the Earth, Parts A/B/C, **31**: 857-863. DOI: 10.1016/j.pce.2006.08.023.

Lennard C, Coop L, Morison D, Grandin R. 2013. Extreme events: Past and future changes in the attributes of extreme rainfall and the dynamics of their driving processes. Climate Systems Analysis Group University of Cape Town.

Li Q, Xing Z, Danielescu S, Li S, Jiang Y, Meng F-R. 2014. Data requirements for using combined conductivity mass balance and recursive digital filter method to estimate groundwater recharge in a small watershed, New Brunswick, Canada. Journal of Hydrology, **511**: 658-664. DOI: http://dx.doi.org/10.1016/j.jhydrol.2014.01.073.

Lindström G, Johansson B, Persson M, Gardelin M, Bergström S. 1997. Development and test of the distributed HBV-96 hydrological model. Journal of Hydrology, **201**: 272-288. DOI: https://doi.org/10.1016/S0022-1694(97)00041-3.

Liu F, Williams MW, Caine N. 2004. Source waters and flow paths in an alpine catchment, Colorado Front Range, United States. Water Resources Research, **40**: W09401. DOI: 10.1029/2004wr003076.

Liu Y, De Smedt F. 2004. WetSpa extension, a GIS-based hydrologic model for flood prediction and watershed management. Vrije Universiteit Brussel, Belgium: 1-108.

Longobardi A, Villani P, Guida D, Cuomo A. 2016. Hydro-geo-chemical streamflow analysis as a support for digital hydrograph filtering in a small, rainfall dominated, sandstone watershed. Journal of Hydrology, **539**: 177-187. DOI: http://dx.doi.org/10.1016/j.jhydrol.2016.05.028.

Lorentz S, Bursey K, Idowu O, Pretorius C, Ngeleka K. 2008. Definition and upscalling of key hydrological processes for application in models. . Water Research Commission, WRC Report No 1320/1/08.

Lorentzen J. 2009. Global sugar, regional water, and local people: EU sugar regime liberalisation, rural livelihoods, and the environment in the Incomati River Basin. South African Journal of Science, **105**: 49-53.

Love D, Uhlenbrook S, Corzo-Perez G, Twomlow S, van der Zaag P. 2010a. Rainfall–interception–evaporation–runoff relationships in a semi-arid catchment, northern Limpopo basin, Zimbabwe. Hydrological Sciences Journal, **55**: 687-703. DOI: 10.1080/02626667.2010.494010.

Love D, Uhlenbrook S, Twomlow S, van Der Zaag P. 2010b. Changing hydroclimatic and discharge patterns in the northern Limpopo Basin, Zimbabwe. Water SA, **36**: 335-350.

Lynch SD. 2003. The Development of a Raster Database of Annual, Monthly and Daily Rainfall for Southern Africa., Water Research Commission, South Africa, Rep. 1156/1/04.

Macamo CC, Balidy H, Bandeira SO, Kairo JG. 2015. Mangrove transformation in the Incomati Estuary, Maputo Bay, Mozambique. Western Indian Ocean Journal of Marine Science, **14**: 11-22.

Maingi JK, Marsh SE. 2002. Quantifying hydrologic impacts following dam construction along the Tana River, Kenya. Journal of Arid Environments, **50**: 53-79. DOI: http://dx.doi.org/10.1006/jare.2000.0860.

Mallory S, Beater A. 2009. Hydrology Report for the Crocodile (East) River Catchment In: Inkomati Water Availability Assessment Study Department of Water Affairs and Forestry, South Africa

Mallory SJ, Hughes DA. 2012. Application of stream flow reduction models within a water resources simulation model. Institute for Water Research, Rhodes University.

Marc V, Didon-Lescot JF, Michael C. 2001. Investigation of the hydrological processes using chemical and isotopic tracers in a small Mediterranean forested catchment during autumn recharge. Journal of Hydrology, **247**: 215-229. DOI: 10.1016/S0022-1694(01)00386-9.

Masih I, Uhlenbrook S, Maskey S, Smakhtin V. 2011. Streamflow trends and climate linkages in the Zagros Mountains, Iran. Climatic Change, **104**: 317-338. DOI: 10.1007/s10584-009-9793-x.

Mathews R, Richter BD. 2007. Application of the Indicators of Hydrologic Alteration Software in Environmental Flow Setting1. JAWRA Journal of the American Water Resources Association, **43**: 1400-1413. DOI: 10.1111/j.1752-1688.2007.00099.x.

Mazvimavi D, Meijerink AMJ, Savenije HHG, Stein A. 2005. Prediction of flow characteristics using multiple regression and neural networks: A case study in Zimbabwe. Physics and Chemistry of the Earth, Parts A/B/C, **30**: 639-647. DOI: http://dx.doi.org/10.1016/j.pce.2005.08.003.

McGlynn BL, McDonnell JJ. 2003. Quantifying the relative contributions of riparian and hillslope zones to catchment runoff. Water Resources Research, **39**: SWC21-SWC220.

McMillan H, Westerberg I, Branger F. 2017. Five guidelines for selecting hydrological signatures. Hydrological Processes, **31**: 4757-4761. DOI: 10.1002/hyp.11300.

Mei Y, Anagnostou EN. 2015. A hydrograph separation method based on information from rainfall and runoff records. Journal of Hydrology, **523**: 636-649. DOI: http://dx.doi.org/10.1016/j.jhydrol.2015.01.083.

Mhlanga BFN, Ndlovu LS, Senzanje A. 2006. Impacts of irrigation return flows on the quality of the receiving waters: A case of sugarcane irrigated fields at the Royal Swaziland Sugar Corporation (RSSC) in the Mbuluzi River Basin (Swaziland). Physics and Chemistry of the Earth, Parts A/B/C, **31**: 804-813.

Miao CY, Shi W, Chen XH, Yang L. 2012. Spatio-temporal variability of streamflow in the Yellow River: possible causes and implications. Hydrological Sciences Journal, **57**: 1355-1367. DOI: 10.1080/02626667.2012.718077.

Middleton BJ, Bailey AK. 2009. Water resources of South Africa, 2005 study. Water Research Commission.

Midgley DC, Pittman WV, Middleton BJ. 1994. Surface Water Resources of South Africa 1990. WRC (ed.) Water Research Commission

Miller MP, Johnson HM, Susong DD, Wolock DM. 2015. A new approach for continuous estimation of baseflow using discrete water quality data: Method description and comparison with baseflow estimates from two existing approaches. Journal of Hydrology, **522**: 203-210. DOI: http://dx.doi.org/10.1016/j.jhydrol.2014.12.039.

Miller MP, Susong DD, Shope CL, Heilweil VM, Stolp BJ. 2014. Continuous estimation of baseflow in snowmelt-dominated streams and rivers in the Upper Colorado River Basin: A chemical hydrograph separation approach. Water Resources Research, **50**: 6986-6999. DOI: 10.1002/2013WR014939.

Milly PCD, Betancourt J, Falkenmark M, Hirsch RM, Kundzewicz ZW, Lettenmaier DP, Stouffer RJ. 2008. Stationarity Is Dead: Whither Water Management? Science, **319**: 573-574. DOI: 10.1126/science.1151915.

Molden D. 1997. Accounting for water use and productivity. International Irrigation Management Institute.

Montanari A, Young G, Savenije HHG, Hughes D, Wagener T, Ren LL, Koutsoyiannis D, Cudennec C, Toth E, Grimaldi S, Blöschl G, Sivapalan M, Beven K, Gupta H, Hipsey M, Schaefli B, Arheimer B, Boegh E, Schymanski SJ, Di Baldassarre G, Yu B, Hubert P, Huang Y, Schumann A, Post DA, Srinivasan V, Harman C, Thompson S, Rogger M, Viglione A, McMillan H, Characklis G, Pang Z, Belyaev V. 2013. "Panta Rhei—Everything Flows": Change in hydrology and society—The IAHS Scientific Decade 2013–2022. Hydrological Sciences Journal, **58**: 1256-1275. DOI: 10.1080/02626667.2013.809088.

Moore D, Dore J, Gyawali D. 2010. The World Commission on Dams+ 10: Revisiting the large dam controversy. Water Alternatives, **3**: 3-13.

Mostert A, McKenzie R, Crerar S. 1993. A rainfall/runoff model for ephemeral rivers in an arid or semi-arid environment. In: 6th South African National Hydrology Symposium, pp: 219-224.

Mukororira F. 2012. Analysis of water allocation in the Komati catchment downstream of Maguga and Driekoppies Dams. In: Water Management, UNESCO-IHE.

Mul ML. 2009. Understanding hydrological processes in an ungauged catchment in sub-Saharan Africa. Technical University of Delft.

Müller C, Hellebrand H, Seeger M, Schobel S. 2009. Identification and regionalization of dominant runoff processes a GIS-based and a statistical approach. Hydrol. Earth Syst. Sci. Hydrology and Earth System Sciences, **13**: 779-792. DOI: 10.5194/hess-13-779-2009.

Muñoz-Villers LE, McDonnell JJ. 2012. Runoff generation in a steep, tropical montane cloud forest catchment on permeable volcanic substrate. Water Resources Research, **48**: W09528. DOI: 10.1029/2011wr011316.

Munyaneza O, Wenninger J, Uhlenbrook S. 2012. Identification of runoff generation processes using hydrometric and tracer methods in a meso-scale catchment in Rwanda. Hydrol. Earth Syst. Sci. Discuss., **9**: 671 - 705. DOI: 10.5194/hessd-9-671-2012.

Mussá F, Zhou Y, Maskey S, Masih I, Uhlenbrook S. 2013. Trend analysis on dry extremes of precipitation and discharge in the Crocodile River catchment, Incomati basin. . In: 14th WaterNet/ WARFSA/GWP-SA symposium, 30 Oct-1 Nov 2013, Dar es Salaam, Tanzania.

Mussa FEF, Zhou Y, Maskey S, Masih I, Uhlenbrook S. 2015. Groundwater as an emergency source for drought mitigation in the Crocodile River catchment, South Africa. Hydrol. Earth Syst. Sci., **19**: 1093-1106. DOI: 10.5194/hess-19-1093-2015.

Nachtergaele F, van Velthuizen H, Verelst L, Batjes N, Dijkshoorn K, van Engelen V, Fischer G, Jones A, Montanarella L, Petri M. 2008. Harmonized world soil database. Food and Agriculture Organization of the United Nations.

Nathan RJ, McMahon TA. 1990. Evaluation of automated techniques for base flow and recession analyses. Water Resources Research, **26**: 1465-1473. DOI: 10.1029/WR026i007p01465.

Nkomo S, van der Zaag P. 2004. Equitable water allocation in a heavily committed international catchment area: the case of the Komati Catchment. Physics and Chemistry of the Earth, Parts A/B/C, **29**: 1309-1317.

O'Brien RJ, Misstear BD, Gill LW, Johnston PM, Flynn R. 2014. Quantifying flows along hydrological pathways by applying a new filtering algorithm in conjunction with master recession curve analysis. Hydrological Processes, **28**: 6211-6221. DOI: 10.1002/hyp.10105.

Paterson G, Turner D, Wiese L, van Zijl G, Clarke C, van Tol J. 2015. Spatial soil information in South Africa: Situational analysis, limitations and challenges. South African Journal of Science, **111**: 1-7.

Pearce AJ, Stewart MK, M.G. S. 1986. Storm runoff generation in humid headwater catchments 1. Where does the water come from? Wat. Resour. Res., **22**: 1263-1273.

Petersen R. 2012. A conceptual understanding of groundwater recharge processes and surface-groundwater interactions in the Kruger National Park In: Department of Earth Sciences, University of the Western Cape.

Pettitt A. 1979. A non-parametric approach to the change-point problem. Appl. Statist., **28 (2)**: 126-135.

Pike A, Schulze R. 1995. AUTOSOILS: A program to convert ISCW soils attributes to variables usable in hydrological models. Pietermaritzburg, South Africa: University of KwaZulu-Natal, School of Bioresources Engineering and Environmental Hydrology.

Pollard S, du Toit D. 2009. Integrated water resource management in complex systems: How the catchment management strategies seek to achieve sustainability and equity in water resources in South Africa. Water SA, **34**: 671-680.

Pollard S, du Toit D. 2011a. Towards Adaptive Integrated Water Resources Management in Southern Africa: The Role of Self-organisation and Multi-scale Feedbacks for Learning and Responsiveness in the Letaba and Crocodile Catchments. Water Resources Management, **25**: 4019-4035. DOI: 10.1007/s11269-011-9904-0.

Pollard S, du Toit D. 2011b. Towards the sustainability of freshwater systems in South Africa: An exploration of factors that enable and constrain meeting the ecological Reserve within the context of Integrated Water Resources Management in the catchments of the lowveld. Water Research Comission.

Pollard S, Du Toit D, Biggs H. 2011. River management under transformation: The emergence of strategic adaptive management of river systems in the Kruger National Park. Koedoe, **53**: Art. #1011. DOI: 10.4102/koedoe.v53i2.1011.

Pollard S, Mallory S, Riddell E, Sawunyama T. 2012. Towards improving the assessment and implementation of the reserve : real-time assessment and implementation of the ecological reserve : report to the Water Research Commission. Water Research Commission.

R Development Core Team. 2014. R: A Language and Environment for Statistical Computing. http://www.R-project.org. Accessed August 1st, 2014

Rennó CD, Nobre AD, Cuartas LA, Soares JV, Hodnett MG, Tomasella J, Waterloo MJ. 2008. HAND, a new terrain descriptor using SRTM-DEM: Mapping terra-firme rainforest environments in Amazonia. Remote Sensing of Environment, **112**: 3469-3481. DOI: http://dx.doi.org/10.1016/j.rse.2008.03.018.

Retief DCH. 2014. Investigating integrated catchment management using a simple water quantity and quality model: A case study of the Crocodile River Catchment, South Africa. Rhodes University, pp: 241.

Richter BD, Baumgartner JV, Braun DP, Powell J. 1998. A spatial assessment of hydrologic alteration within a river network. Regulated Rivers: Research & Management, **14**: 329-340. DOI: 10.1002/(sici)1099-1646(199807/08)14:4<329::aid-rrr505>3.0.co;2-e.

Richter BD, Baumgartner JV, Powell J, Braun DP. 1996. A Method for Assessing Hydrologic Alteration within Ecosystems. Conservation Biology, **10**: 1163-1174.

Richter BD, Mathews R, Harrison DL, Wigington R. 2003. Ecologically sustainable water management: managing river flows for ecological integrity. Ecological Applications, **13**: 206-224. DOI: 10.1890/1051-0761(2003)013[0206:eswmmr]2.0.co;2.

Richter BD, Thomas GA. 2007. Restoring environmental flows by modifying dam operations. Ecology and Society, **12**: 12.

Riddell E, Everson C, Clulow A, Mengistu M. 2013. The hydrological characterisation and water budget of a South African rehabilitated headwater wetland system. Water SA, **39**: 57-66.

Riddell E, Jewitt G. 2010. Report providing updated catchment information and identifying study focus areas. A Management Tool for the Inkomati Basin with focus on Improved Hydrological Understanding for Risk-based Operational Water Management., Water Research Comission, Project K5/1935.

Riddell E, Jewitt G, Chetty K, Saraiva Okello A, Jackson B, Lamba A, Gokoo S, Naidoo P, Vather T, Thornton-Dibb S. 2014a. A management tool for the Inkomati Basin with focus on improved hydrological understanding for risk-based operational water management. Riddell E, Jewitt G (eds.) WRC.

Riddell E, Pollard S, Mallory S, Sawunyama T. 2014b. A methodology for historical assessment of compliance with environmental water allocations: lessons from the Crocodile (East) River, South Africa. Hydrological Sciences Journal, **59**: 831-843. DOI: 10.1080/02626667.2013.853123.

Rijsberman FR. 2006. Water scarcity: Fact or fiction? Agricultural Water Management, **80**: 5-22.

Rimmer A, Hartmann A. 2014. Optimal hydrograph separation filter to evaluate transport routines of hydrological models. Journal of Hydrology, **514**: 249-257. DOI: http://dx.doi.org/10.1016/j.jhydrol.2014.04.033.

Rockström J, Falkenmark M, Karlberg L, Hoff H, Rost S, Gerten D. 2009. Future water availability for global food production: the potential of green water for increasing resilience to global change. Water Resources Research, **45**.

Rockström J, Folke C, Gordon L, Hatibu N, Jewitt G, Penning de Vries F, Rwehumbiza F, Sally H, Savenije H, Schulze R. 2004. A watershed approach to upgrade rainfed agriculture in water scarce regions through Water System Innovations: an integrated research initiative on water for food and rural livelihoods in balance with ecosystem functions. Physics and Chemistry of the Earth, Parts A/B/C, **29**: 1109-1118. DOI: http://dx.doi.org/10.1016/j.pce.2004.09.016.

Rouault M, Fauchereau N, Pohl B, Penven P, Richard Y, Reason C, Pegram G, Phillippon N, Siedler G, Murgia A. 2010. Multidisciplinary analysis of hydroclimatic variability at the catchment scale. University of Cape Town; CRC, Universite de Dijon; UBO,Universite de Bretagne Occidentale; University of Kwazulu Natal; IMF, University of Kiel; LTHE, Universite de Grenoble.

Sahula A. 2014. Exploring the development of an integrated, participative, water quality management process for the Crocodile River Catchment, focusing on the sugar industry. MSc Thesis, Rhodes University, pp: 265.

Saraiva Okello AML, Masih I, Uhlenbrook S, Jewitt GPW, Van der Zaag P. 2018a. Improved Process Representation in the Simulation of the Hydrology of a Meso-Scale Semi-Arid Catchment. Water, **10**: 1549. DOI: https://doi.org/10.3390/w10111549.

Saraiva Okello AML, Masih I, Uhlenbrook S, Jewitt GPW, van der Zaag P, Riddell E. 2015. Drivers of spatial and temporal variability of streamflow in the Incomati River basin. Hydrol. Earth Syst. Sci., **19**: 657-673. DOI: 10.5194/hess-19-657-2015.

Saraiva Okello AML, Uhlenbrook S, Jewitt GPW, Masih I, Riddell ES, Van der Zaag P. 2018b. Hydrograph separation using tracers and digital filters to quantify runoff components in a semi-arid mesoscale catchment. Hydrological Processes, **32**: 1334-1350. DOI: 10.1002/hyp.11491.

Savenije HHG. 1995. Spreadsheets: flexible tools for integrated management of water resources in river basins. IAHS PUBLICATION: 207-216.

Savenije HHG. 2000. Foreword Water scarcity, water conservation and water resources valuation. Physics and Chemistry of the Earth, Part B: Hydrology, Oceans and Atmosphere, **25**: 191-191.

Savenije HHG. 2010. HESS Opinions "Topography driven conceptual modelling (FLEX-Topo)". Hydrol. Earth Syst. Sci., **14**: 2681-2692. DOI: 10.5194/hess-14-2681-2010.

Savenije HHG, van der Zaag P. 2000. Conceptual framework for the management of shared river basins; with special reference to the SADC and EU. Water Policy, **2**: 9-45.

Scherrer S, Naef F. 2003. A decision scheme to indicate dominant hydrological flow processes on temperate grassland. Hydrological Processes, **17**: 391-401. DOI: 10.1002/hyp.1131.

Schulze R. 1985. Hydrological characteristics and properties of soils in Southern Africa 1: Runoff response. Water SA, **11**: 121-128.

Schulze R. 2012. A 2011 perspective on climate change and the South African water sector. WRC Report TT 518/12.

Schulze R, Maharaj M, Warburton M, Gers C, Horan M, Kunz R, Clark D. 2007. South African atlas of climatology and agrohydrology. Water Research Commission, Pretoria, RSA, WRC Report, **1489**: 06.

Schulze RE. 1997. South African atlas of agrohydrology and -climatology. Water Research Commission.

Schulze RE. 2000. Modelling Hydrological Responses to Land Use and Climate Change: A Southern African Perspective. AMBIO: A Journal of the Human Environment, **29**: 12-22. DOI: 10.1579/0044-7447-29.1.12.

Schulze RE. 2011. Approaches towards practical adaptive management options for selected water-related sectors in South Africa in a context of climate change. Water SA, **37**: 621-645.

Scott DF, Lesch W. 1997. Streamflow responses to afforestation with Eucalyptus grandis and Pinus patula and to felling in the Mokobulaan experimental catchments, South Africa. Journal of Hydrology, **199**: 360-377. DOI: https://doi.org/10.1016/S0022-1694(96)03336-7.

Senay GB, Bohms S, Singh RK, Gowda PH, Velpuri NM, Alemu H, Verdin JP. 2013. Operational evapotranspiration mapping using remote sensing and weather datasets: A new parameterization for the SSEB approach. JAWRA Journal of the American Water Resources Association, **49**: 577-591. DOI: 10.1111/jawr.12057.

Sengo DJ, Kachapila A, van der Zaag P, Mul M, Nkomo S. 2005. Valuing environmental water pulses into the Incomati estuary: Key to achieving equitable and sustainable utilisation of transboundary waters. Physics and Chemistry of the Earth, Parts A/B/C, **30**: 648-657.

Sharpe MR, Sohnge AP, Zyl JSv, Joubert DK, Mulder MP, Clubley-Armstrong AR, Plessis CPd, Eeden ORv, Rossouw PJ, Visser DJL, Viljoen MJ, Taljaard JJ. 1986. 2530 Barberton. Geoscience Cf (ed.) Department of Minerals and Energy Affairs, Government Printer.

Shongwe ME, van Oldenborgh GJ, van den Hurk BJJM, de Boer B, Coelho CAS, van Aalst MK. 2009. Projected Changes in Mean and Extreme Precipitation in Africa under Global Warming. Part I: Southern Africa. Journal of Climate, **22**: 3819-3837. DOI: 10.1175/2009jcli2317.1.

Sieber A, Uhlenbrook S. 2005. Sensitivity analyses of a distributed catchment model to verify the model structure. Journal of Hydrology, **310**: 216-235. DOI: http://dx.doi.org/10.1016/j.jhydrol.2005.01.004.

Sklash MG, Farvolden RN. 1979. The role of groundwater in storm runoff. Journal of Hydrology, **43**: 45-65. DOI: http://dx.doi.org/10.1016/0022-1694(79)90164-1.

Slaughter AR, Hughes DA. 2013. A simple model to separately simulate point and diffuse nutrient signatures in stream flows. Hydrology Research, **44**: 538-553.

Slinger JH, Hilders M, Juizo D. 2010. The practice of transboundary decision-making on the Incomati River: elucidating underlying factors and their implications for institutional design. Ecology and Society, **15**: 1.

Smith RE, Goodrich DC. 2005. Rainfall excess overland flow. In: Encyclopedia of Hydrological Sciences, Anderson MG (ed.) John Wiley & Sons, Ltd, pp: 1707-1718.

Smithers JC, Schulze RE, Pike A, Jewitt GPW. 2001. A hydrological perspective of the February 2000 floods: a case study in the Sabie River catchment. Water SA, **27**: 325-332.

Soulsby C, Rodgers PJ, Petry J, Hannah DM, Malcolm IA, Dunn SM. 2004. Using tracers to upscale flow path understanding in mesoscale mountainous catchments: two examples from Scotland. Journal of Hydrology, **291**: 174-196. DOI: http://dx.doi.org/10.1016/j.jhydrol.2003.12.042.

Soulsby C, Tetzlaff D, Rodgers P, Dunn S, Waldron S. 2006. Runoff processes, stream water residence times and controlling landscape characteristics in a mesoscale catchment: An initial evaluation. Journal of Hydrology, **325**: 197-221. DOI: http://dx.doi.org/10.1016/j.jhydrol.2005.10.024.

Stats S. 2016. Mid-year population estimates 2016. Statistics South Africa.

Stewart MK. 2015. Promising new baseflow separation and recession analysis methods applied to streamflow at Glendhu Catchment, New Zealand. Hydrol. Earth Syst. Sci., **19**: 2587-2603. DOI: 10.5194/hess-19-2587-2015.

Tallaksen LM. 1995. A review of baseflow recession analysis. Journal of Hydrology, **165**: 349-370. DOI: http://dx.doi.org/10.1016/0022-1694(94)02540-R.

Taylor V, Schulze R, Jewitt GPW. 2003. Application of the Indicators of Hydrological Alteration method to the Mkomazi River, KwaZulu-Natal, South Africa. African Journal of Aquatic Science, **28**: 1-11. DOI: 10.2989/16085914.2003.9626593.

Tetzlaff D, Buttle J, Carey SK, McGuire K, Laudon H, Soulsby C. 2015. Tracer-based assessment of flow paths, storage and runoff generation in northern catchments: a review. Hydrological Processes, **29**: 3475-3490. DOI: 10.1002/hyp.10412.

Tetzlaff D, Soulsby C. 2008. Sources of baseflow in larger catchments – Using tracers to develop a holistic understanding of runoff generation. Journal of Hydrology, **359**: 287-302. DOI: http://dx.doi.org/10.1016/j.jhydrol.2008.07.008.

Tetzlaff D, Waldron S, Brewer MJ, Soulsby C. 2007. Assessing nested hydrological and hydrochemical behaviour of a mesoscale catchment using continuous tracer data. Journal of Hydrology, **336**: 430-443.

TIA. 2002. Tripartite Interim Agreement between Mozambique, South Africa and Swaziland for co-operation on the protection and sustainable utilisation of the water resources of the Incomati and Maputo watercourses. Johannesburg, 29 August 2002.

TPTC. 2010. Tripartite Permanent Technical Committee (TPTC) between, Moçambique, South Africa, Swaziland, 2010. PRIMA : IAAP 3 : Consultancy Services for Integrated Water Resources Management. Baseline evaluation and scoping report: Part C. Report No.:IAAP 3: 03C - 2010. Prepared for TPTC by Aurecon. Aurecon PfTb (ed.).

Trambauer P, Maskey S, Winsemius H, Werner M, Uhlenbrook S. 2013. A review of continental scale hydrological models and their suitability for drought forecasting in (sub-Saharan) Africa. Physics and Chemistry of the Earth, Parts A/B/C, **66**: 16-26. DOI: http://dx.doi.org/10.1016/j.pce.2013.07.003.

Tullos D, Tilt B, Liermann CR. 2009. Introduction to the special issue: Understanding and linking the biophysical, socioeconomic and geopolitical effects of dams. Journal of Environmental Management, **90, Supplement 3**: S203-S207. DOI: http://dx.doi.org/10.1016/j.jenvman.2008.08.018.

Uhlenbrook S. 2003. An empirical approach for delineating spatial units with the same dominating runoff generation processes. Physics and Chemistry of the Earth, Parts A/B/C, **28**: 297-303.

Uhlenbrook S. 2006. Catchment hydrology—a science in which all processes are preferential. Hydrological Processes, **20**: 3581-3585. DOI: 10.1002/hyp.6564.

Uhlenbrook S. 2009. Climate and man-made changes and their impacts on catchments. In: Water Policy 2009, Water as a Vulnerable and Exhaustible Resource. Proceedings of the Joint Conference of APLU and ICA, 23-26 June 2009, Prague, Czech Republic, page 81-87., Kovar P, Maca P, Redinova J (eds.).

Uhlenbrook S, Frey M, Leibundgut C, Maloszewski P. 2002. Hydrograph separations in a mesoscale mountainous basin at event and seasonal timescales. Water Resources Research, **38**: 31-31-31-14. DOI: 10.1029/2001wr000938.

Uhlenbrook S, Hoeg S. 2003. Quantifying uncertainties in tracer-based hydrograph separations: a case study for two-, three- and five-component hydrograph separations in a mountainous catchment. Hydrological Processes, **17**: 431-453. DOI: 10.1002/hyp.1134.

Uhlenbrook S, Leibundgut C. 2000. Development and validation of a process oriented catchment model based on dominating runoff generation processes. Physics and Chemistry of the Earth, Part B: Hydrology, Oceans and Atmosphere, **25**: 653-657.

Uhlenbrook S, Roser S, Tilch N. 2004. Hydrological process representation at the meso-scale: the potential of a distributed, conceptual catchment model. Journal of Hydrology, **291**: 278-296.

UNEP. 1997. World atlas of desertification 2ED. United Nations Environment Programme (UNEP).

Van den Berg E, Plarre C, Van den Berg H, Thompson M. 2008. The South African national land cover 2000. Agricultural Research Council (ARC) and Council for Scientific and Industrial Research (CSIR), Pretoria. Report No. GW/A/2008/86.

Van der Zaag P, Carmo Vaz A. 2003. Sharing the Incomati water:cooperation and competition in the balance. Water Policy, **5**: 346-368.

van der Zaag P, Seyam IM, Savenije HHG. 2002. Towards measurable criteria for the equitable sharing of international water resources. Water Policy, **4**: 19-32. DOI: http://dx.doi.org/10.1016/S1366-7017(02)00003-X.

Van der Zaag P, Vaz AC. 2003. Sharing the Incomati waters: cooperation and competition in the balance. Water Policy, **5**: 349-368.

van Eekelen MW, Bastiaanssen WGM, Jarmain C, Jackson B, Ferreira F, van der Zaag P, Saraiva Okello A, Bosch J, Dye P, Bastidas-Obando E, Dost RJJ, Luxemburg WMJ. 2015. A novel approach to estimate direct and indirect water withdrawals from satellite measurements: A case study from the Incomati basin. Agriculture, Ecosystems & Environment, **200**: 126-142. DOI: http://dx.doi.org/10.1016/j.agee.2014.10.023.

van Engelen V, Dijkshoorn J. 2013. Global and National Soils and Terrain Digital Databases (SOTER)-Procedures manual (Ver. 2.0). ISRIC Report, **4**.

Van Huyssteen CW. 2008. A review of advances in hydropedology for application in South Africa : review article. South African Journal of Plant and Soil, **25**: 245-253.

van Tol JJ, Le Roux PAL, Lorentz SA, Hensley M. 2013. Hydropedological Classification of South African Hillslopes. Vadose Zone Journal, **12**. DOI: 10.2136/vzj2013.01.0007.

Van Tol JJ, Van Zijl GM, Riddell ES, Fundisi D. 2015. Application of hydropedological insights in hydrological modelling of the Stevenson-Hamilton Research Supersite, Kruger National Park, South Africa. Water SA, **41**: 525-533. DOI: doi:http://dx.doi.org/10.4314/wsa.v41i4.12.

van Wyk E, van Tonder G, Vermeulen D. 2011. Characteristics of local groundwater recharge cycles in South African semi-arid hard rock terrains - rainwater input. Water SA, **37**: 147-154.

Van Wyk E, Van Tonder G, Vermeulen D. 2012. Characteristics of local groundwater recharge cycles in South African semi-arid hard rock terrains: Rainfall-groundwater interaction. Water SA, **38**: 747-754.

van Zijl G, Le Roux P. 2014. Creating a conceptual hydrological soil response map for the Stevenson Hamilton Research Supersite, Kruger National Park, South Africa. Water SA, **40**: 331-336.

van Zijl GM, van Tol JJ, Riddell ES. 2016. Digital soil mapping for hydrological modelling. In: Digital Soil Mapping Across Paradigms, Scales and Boundaries, Zhang G-L, Brus D, Liu F, Song X-D, Lagacherie P (eds.) Springer Singapore, pp: 115-129.

Viglione A, Parajka J, Rogger M, Salinas JL, Laaha G, Sivapalan M, Blöschl G. 2013. Comparative assessment of predictions in ungauged basins – Part 3: Runoff signatures in Austria. Hydrol. Earth Syst. Sci., **17**: 2263-2279. DOI: 10.5194/hess-17-2263-2013.

Vörösmarty CJ, McIntyre P, Gessner MO, Dudgeon D, Prusevich A, Green P, Glidden S, Bunn SE, Sullivan CA, Liermann CR. 2010. Global threats to human water security and river biodiversity. Nature, **467**: 555-561.

Wang R, Kumar M, Marks D. 2013. Anomalous trend in soil evaporation in a semi-arid, snow-dominated watershed. Advances in Water Resources: 32-40.

Warburton ML, Schulze RE, Jewitt GPW. 2010. Confirmation of ACRU model results for applications in land use and climate change studies. Hydrol. Earth Syst. Sci., **14**: 2399-2414. DOI: 10.5194/hess-14-2399-2010.

Warburton ML, Schulze RE, Jewitt GPW. 2012. Hydrological impacts of land use change in three diverse South African catchments. Journal of Hydrology, **414–415**: 118-135. DOI: http://dx.doi.org/10.1016/j.jhydrol.2011.10.028.

Ward PJ, Aerts JCJH, de Moel H, Renssen H. 2007. Verification of a coupled climate-hydrological model against Holocene palaeohydrological records. Global and Planetary Change, **57**: 283-300. DOI: https://doi.org/10.1016/j.gloplacha.2006.12.002.

Ward PJ, Renssen H, Aerts J, Van Balen R, Vandenberghe J. 2008. Strong increases in flood frequency and discharge of the River Meuse over the late Holocene: impacts of long-term anthropogenic land use change and climate variability. Hydrol. Earth Syst. Sci., **12**: 159-175. DOI: 10.5194/hess-12-159-2008.

Ward PJ, Renssen H, Aerts JCJH, Verburg PH. 2011. Sensitivity of discharge and flood frequency to twenty-first century and late Holocene changes in climate and land use (River Meuse, northwest Europe). Climatic Change, **106**: 179-202. DOI: 10.1007/s10584-010-9926-2.

Weiler M, McGlynn BL, McGuire KJ, McDonnell JJ. 2003. How does rainfall become runoff? A combined tracer and runoff transfer function approach. Water Resources Research, **39**.

Wenninger J, Uhlenbrook S, Lorentz S, Leibundgut C. 2008. Idenfication of runoff generation processes using combined hydrometric, tracer and geophysical methods in a headwater catchment in South Africa. Hydrological Sciences Journal, **53**: 65-80. DOI: 10.1623/hysj.53.1.65.

Wheater H, Sorooshian S, Sharma KD. 2008. Hydrological modelling in arid and semi-arid areas. Cambridge University Press.

Winsemius HC, Savenije HHG, Gerrits AMJ, Zapreeva EA, Klees R. 2006. Comparison of two model approaches in the Zambezi river basin with regard to model reliability and identifiability. Hydrol. Earth Syst. Sci., **10**: 339-352. DOI: 10.5194/hess-10-339-2006.

Winston WE, Criss RE. 2002. Geochemical variations during flash flooding, Meramec River basin, May 2000. Journal of Hydrology, **265**: 149-163. DOI: 10.1016/S0022-1694(02)00105-1.

Wissmeier L, Uhlenbrook S. 2007. Distributed, high-resolution modelling of 18O signals in a meso-scale catchment. Journal of Hydrology, **332**: 497-510.

Yepez EA, Williams DG, Scott RL, Lin G. 2003. Partitioning overstory and understory evapotranspiration in a semiarid savanna woodland from the isotopic composition of water vapor. Agricultural and Forest Meteorology, **119**: 53-68.

Zhang R, Li Q, Chow TL, Li S, Danielescu S. 2013. Baseflow separation in a small watershed in New Brunswick, Canada, using a recursive digital filter calibrated with the conductivity mass balance method. Hydrological Processes, **27**: 2659-2665. DOI: 10.1002/hyp.9417.

Zhang X, Zhang L, Zhao J, Rustomji P, Hairsine P. 2008. Responses of streamflow to changes in climate and land use/cover in the Loess Plateau, China. Water Resources Research, **44**: W00A07. DOI: 10.1029/2007wr006711.

Appendices

A1 (Chapter 5) – Water quality statistics

Data for the analysis was obtained from Water Management System https://www.dwa.gov.za/iwqs/wms/

Descriptive Statistics

X2H008 - Queens	*Ca*	*Cl*	*DMS*	*EC*	*F*	*K*	*Mg*	*Na*	*NH4_N*	*NO3_NO2_N*	*pH*	*PO4_P*	*Si*	*SO4*	*TAL*
Mean	**10.81**	**5.15**	**127.34**	**16.73**	**0.13**	**0.60**	**10.07**	**7.40**	**0.04**	**0.09**	**7.61**	**0.02**	**11.82**	**6.81**	**70.13**
Standard Deviation	**2.63**	**2.98**	**25.38**	**3.49**	**0.07**	**0.44**	**2.51**	**2.49**	**0.03**	**0.10**	**0.47**	**0.07**	**2.29**	**3.93**	**14.84**
Coefficient of Variation CV	0.24	0.58	0.20	0.21	0.54	0.73	0.25	0.34	0.73	1.05	0.06	3.07	0.19	0.58	0.21
Range	**26.60**	**43.51**	**217.62**	**33.90**	**0.53**	**3.55**	**16.27**	**21.98**	**0.16**	**0.87**	**2.82**	**1.26**	**21.02**	**33.91**	**93.73**
Minimum	4.70	1.50	44.73	5.80	0.03	0.06	2.70	1.00	0.02	0.01	6.24	0.00	2.67	2.00	21.27
Maximum	31.30	45.01	262.36	39.70	0.55	3.61	18.98	22.98	0.18	0.88	9.06	1.26	23.69	35.91	115.00
Count	403	402	391	529	395	404	401	404	402	403	406	401	402	401	403
Confidence Level(95.0%)	0.26	0.29	2.52	0.30	0.01	0.04	0.25	0.24	0.00	0.01	0.05	0.01	0.22	0.39	1.45

X2H010 - Noordkaap	*Ca*	*Cl*	*DMS*	*EC*	*F*	*K*	*Mg*	*Na*	*NH4_N*	*NO3_NO2_N*	*pH*	*PO4_P*	*Si*	*SO4*	*TAL*
Mean	**7.05**	**4.73**	**80.61**	**10.36**	**0.13**	**0.65**	**3.95**	**7.53**	**0.03**	**0.05**	**7.43**	**0.02**	**12.22**	**4.20**	**42.77**
Standard Deviation	**2.04**	**2.36**	**18.32**	**2.52**	**0.08**	**0.45**	**1.24**	**1.92**	**0.03**	**0.05**	**0.59**	**0.02**	**2.17**	**3.13**	**10.63**
Coefficient of Variation CV	0.29	0.50	0.23	0.24	0.62	0.69	0.31	0.25	0.89	0.96	0.08	0.95	0.18	0.75	0.25
Range	**20**	**28.823**	**157**	**25.5**	**0.5**	**2.72**	**12.501**	**14.027**	**0.312**	**0.474**	**3.65**	**0.147**	**22.298**	**24.679**	**100**
Minimum	1.3	1.5	37	4.1	0.05	0.15	1.8	1	0.015	0.005	4.81	0.003	1.572	1.233	17.7
Maximum	21.3	30.323	194	29.6	0.55	2.87	14.301	15.027	0.327	0.479	8.46	0.15	23.87	25.912	117.7
Count	367	367	357	456	361	368	367	367	366	366	371	364	366	366	368
Confidence Level(95.0%)	0.21	0.24	1.91	0.23	0.01	0.05	0.13	0.20	0.00	0.00	0.06	0.00	0.22	0.32	1.09

X2H031 - Suidkaap	*Ca*	*Cl*	*DMS*	*EC*	*F*	*K*	*Mg*	*Na*	*NH4_N*	*NO3_NO2_N*	*pH*	*PO4_P*	*Si*	*SO4*	*TAL*
Mean	**10.00**	**6.14**	**115.03**	**14.64**	**0.17**	**0.83**	**6.12**	**10.55**	**0.04**	**0.13**	**7.59**	**0.04**	**12.32**	**5.56**	**61.24**
Standard Deviation	**2.86**	**3.22**	**25.16**	**3.78**	**0.07**	**0.60**	**1.66**	**2.73**	**0.03**	**0.11**	**0.54**	**0.32**	**2.24**	**3.88**	**13.92**
Coefficient of Variation CV	0.29	0.52	0.22	0.26	0.42	0.72	0.27	0.26	0.81	0.82	0.07	8.49	0.18	0.70	0.23
Range	**28.00**	**48.60**	**217.67**	**33.90**	**0.57**	**4.80**	**16.23**	**22.00**	**0.26**	**0.51**	**3.29**	**6.24**	**24.74**	**34.41**	**94.10**
Minimum	0.50	1.50	50.00	6.00	0.03	0.15	1.59	1.00	0.02	0.02	5.89	0.00	0.20	1.50	17.90
Maximum	28.50	50.10	267.67	39.90	0.59	4.95	17.81	23.00	0.28	0.53	9.18	6.25	24.94	35.91	112.00
Count	372	373	355	527	363	369	372	370	370	372	375	370	370	373	371
Confidence Level(95.0%)	0.29	0.33	2.63	0.32	0.01	0.06	0.17	0.28	0.00	0.01	0.05	0.03	0.23	0.39	1.42

X2H022 - Kaap	*Ca*	*Cl*	*DMS*	*EC*	*F*	*K*	*Mg*	*Na*	*NH4_N*	*NO3_NO2_N*	*pH*	*PO4_P*	*Si*	*SO4*	*TAL*
Mean	**29.16**	**22.48**	**429.02**	**53.21**	**0.37**	**1.27**	**31.48**	**42.30**	**0.06**	**0.57**	**8.11**	**0.03**	**13.99**	**50.81**	**204.13**
Standard Deviation	**8.55**	**9.60**	**160.36**	**18.17**	**0.17**	**0.73**	**12.39**	**21.96**	**0.06**	**0.48**	**0.40**	**0.02**	**2.94**	**28.44**	**77.89**
Coefficient of Variation CV	0.29	0.43	0.37	0.34	0.46	0.57	0.39	0.52	1.00	0.85	0.05	0.87	0.21	0.56	0.38
Range	**52.60**	**53.54**	**766.08**	**98.02**	**0.96**	**5.49**	**59.83**	**101.18**	**0.65**	**6.71**	**2.83**	**0.18**	**24.23**	**248.80**	**476.73**
Minimum	8.00	3.70	87.00	11.98	0.05	0.15	4.70	6.25	0.02	0.02	6.41	0.00	1.88	2.00	33.52
Maximum	60.60	57.24	853.08	110.00	1.01	5.64	64.53	107.43	0.66	6.73	9.24	0.18	26.11	250.80	510.25
Count	535.00	536.00	523.00	984.00	526.00	533.00	535.00	535.00	534.00	536.00	542.00	534.00	534.00	533.00	537.00
Confidence Level(95.0%)	0.73	0.81	13.78	1.14	0.01	0.06	1.05	1.86	0.00	0.04	0.03	0.00	0.25	2.42	6.60

Correlation matrix

X2H008	*Ca*	*Cl*	*DMS*	*EC*	*F*	*K*	*Mg*	*Na*	*NH4_N*	*'O3_NO2_I*	*pH*	*PO4_P*	*Si*	*SO4*	*TAL*
Ca	1														
Cl	0.31	1.00													
DMS	0.86	0.35	1.00												
EC	0.82	0.40	0.91	1.00											
F	0.26	0.10	0.32	0.27	1.00										
K	0.06	0.21	0.10	0.10	0.25	1.00									
Mg	0.54	0.20	0.75	0.73	0.03	-0.06	1.00								
Na	0.47	0.32	0.54	0.53	0.31	0.16	0.13	1.00							
NH4_N	-0.09	-0.13	-0.09	-0.06	-0.09	-0.13	0.07	-0.10	1.00						
NO3_NO2	0.08	0.23	0.06	0.07	0.10	0.05	-0.08	0.15	-0.06	1.00					
pH	0.22	0.20	0.26	0.22	0.17	0.20	0.04	0.00	-0.10	-0.03	1.00				
PO4_P	0.07	0.02	0.00	-0.01	0.12	0.04	-0.01	0.02	-0.02	0.38	-0.07	1.00			
Si	0.39	-0.12	0.49	0.44	0.24	-0.09	0.38	0.51	0.01	0.10	-0.19	-0.01	1.00		
SO4	0.28	0.42	0.37	0.34	0.11	0.24	0.25	0.11	-0.23	0.05	0.24	0.05	-0.17	1.00	
TAL	0.80	0.11	0.95	0.82	0.30	0.01	0.72	0.44	-0.03	-0.08	0.23	-0.05	0.56	0.14	1

X2H010	*Ca*	*Cl*	*DMS*	*EC*	*F*	*K*	*Mg*	*Na*	*NH4_N*	*'O3_NO2_I*	*pH*	*PO4_P*	*Si*	*SO4*	*TAL*
Ca	1.00														
Cl	0.35	1.00													
DMS	0.86	0.38	1.00												
EC	0.84	0.43	0.91	1.00											
F	0.32	0.17	0.46	0.39	1.00										
K	0.05	0.18	0.18	0.17	0.25	1.00									
Mg	0.83	0.34	0.82	0.84	0.24	0.06	1.00								
Na	0.24	0.20	0.51	0.46	0.27	0.13	0.23	1.00							
NH4_N	-0.01	0.05	0.01	0.06	-0.01	0.00	0.03	0.05	1.00						
NO3_NO2	0.18	0.23	0.13	0.21	-0.02	0.06	0.25	-0.18	0.08	1.00					
pH	0.49	0.30	0.54	0.49	0.41	0.10	0.40	0.12	-0.07	0.20	1.00				
PO4_P	0.03	0.01	0.04	0.03	0.02	0.10	-0.01	-0.01	0.02	0.20	0.18	1.00			
Si	-0.07	-0.18	0.10	0.09	0.05	-0.12	-0.06	0.56	-0.04	-0.38	-0.11	-0.02	1.00		
SO4	0.19	0.31	0.33	0.17	0.15	0.13	0.21	0.02	-0.02	0.20	0.12	0.04	-0.22	1.00	
TAL	0.80	0.15	0.95	0.84	0.44	0.11	0.77	0.44	-0.01	0.04	0.52	0.04	0.17	0.09	1.00

X2H031	*Ca*	*Cl*	*DMS*	*EC*	*F*	*K*	*Mg*	*Na*	*NH4_N*	*!O3_NO2_I*	*pH*	*PO4_P*	*Si*	*SO4*	*TAL*
Ca	1.00														
Cl	0.46	1.00													
DMS	0.85	0.57	1.00												
EC	0.85	0.58	0.93	1.00											
F	0.30	0.17	0.40	0.38	1.00										
K	-0.05	0.19	0.09	0.09	0.19	1.00									
Mg	0.74	0.59	0.89	0.85	0.34	-0.02	1.00								
Na	0.43	0.37	0.74	0.64	0.30	0.12	0.61	1.00							
NH4_N	0.03	0.10	0.06	0.11	0.08	0.00	0.10	0.05	1.00						
NO3_NO2	0.12	0.13	0.15	0.20	0.07	0.19	0.11	0.01	0.05	1.00					
pH	0.39	0.26	0.45	0.45	0.29	0.20	0.38	0.12	-0.02	0.42	1.00				
PO4_P	0.02	0.05	0.04	0.05	-0.01	0.05	0.07	0.00	-0.02	-0.02	0.06	1.00			
Si	0.01	-0.11	0.08	0.09	0.02	-0.13	0.09	0.29	0.03	0.00	-0.10	-0.05	1.00		
SO4	0.28	0.42	0.38	0.32	0.19	0.33	0.29	0.22	0.04	0.20	0.22	0.01	-0.21	1.00	
TAL	0.80	0.33	0.94	0.85	0.36	-0.06	0.81	0.63	0.04	0.09	0.41	0.11	0.17	0.14	1.00

X2H022	*Ca*	*Cl*	*DMS*	*EC*	*F*	*K*	*Mg*	*Na*	*NH4_N*	*!O3_NO2_I*	*pH*	*PO4_P*	*Si*	*SO4*	*TAL*
Ca	1														
Cl	0.81	1.00													
DMS	0.90	0.91	1.00												
EC	0.91	0.90	0.99	1.00											
F	0.50	0.58	0.71	0.68	1.00										
K	-0.06	0.04	-0.14	-0.12	-0.16	1.00									
Mg	0.90	0.89	0.96	0.96	0.59	-0.14	1.00								
Na	0.72	0.84	0.93	0.90	0.82	-0.13	0.81	1.00							
NH4_N	0.08	0.06	0.09	0.09	-0.06	-0.11	0.09	0.06	1.00						
NO3_NO2	0.51	0.41	0.47	0.49	0.29	-0.16	0.49	0.38	0.03	1.00					
pH	0.40	0.45	0.41	0.42	0.16	-0.02	0.43	0.30	-0.06	0.34	1.00				
PO4_P	-0.04	-0.02	-0.05	-0.05	-0.01	0.03	-0.08	-0.04	0.04	0.03	0.05	1.00			
Si	0.38	0.31	0.46	0.45	0.44	-0.29	0.50	0.40	0.13	0.27	0.04	-0.07	1.00		
SO4	0.77	0.73	0.71	0.74	0.35	0.28	0.71	0.61	0.04	0.24	0.32	-0.08	0.05	1.00	
TAL	0.83	0.85	0.97	0.94	0.73	-0.28	0.93	0.90	0.10	0.48	0.39	-0.04	0.55	0.55	1.00

A2 (Chapter 5) – Monthly and Annual Hydrograph Separation Results

Monthly Results Kaap

	Flow mean	Flow std	Flow min	Flow max	bf mean	bf std	bf min	bf max	qf mean	qf std	qf min	qf max	BFI mean	BFI std	BFI min	BFI max	qf/bf
Month																	
January	15.91	16.58	0.27	58.01	4.64	4.89	0.06	18.48	11.28	13.21	0.11	46.14	0.34	0.19	0.09	0.78	2.43
February	22.74	46.81	0.02	244.83	6.63	10.14	0.01	51.16	**16.10**	37.03	0.01	193.67	0.38	0.18	0.13	0.78	2.43
March	16.76	22.18	0.05	98.82	**7.34**	11.33	0.03	54.51	9.42	11.81	0.02	44.30	0.46	0.21	0.16	0.89	1.28
April	10.32	14.75	0.03	77.43	5.52	7.81	0.03	43.60	4.80	7.94	0.01	33.83	0.55	0.25	0.15	0.97	0.87
May	5.46	7.43	0.03	39.85	3.78	5.37	0.02	29.50	1.68	2.20	0.01	10.35	0.64	0.19	0.16	0.98	0.44
June	3.53	4.18	0.01	21.07	2.33	3.20	0.00	17.07	1.20	1.20	0.01	3.99	0.59	0.16	0.05	0.82	0.51
July	2.78	3.00	0.00	14.45	1.68	2.09	0.00	10.91	1.10	1.05	0.00	3.54	0.55	0.15	0.09	0.87	0.66
August	2.04	2.21	0.00	8.81	1.19	1.39	0.00	6.69	**0.84**	0.91	0.00	3.39	0.56	0.15	0.23	0.87	0.71
September	1.82	2.27	0.00	8.87	0.86	0.98	0.00	4.10	0.96	1.55	0.00	6.80	0.52	0.22	0.09	0.84	1.11
October	2.36	2.91	0.00	11.58	**0.77**	0.87	0.00	3.22	1.59	2.24	0.00	9.08	0.33	0.18	0.03	0.73	2.07
November	6.42	8.36	0.00	33.34	1.36	1.52	0.00	6.71	5.06	7.00	0.00	26.63	0.25	0.14	0.05	0.61	3.71
December	12.63	15.23	0.02	66.05	3.25	3.98	0.01	15.03	9.38	11.43	0.01	51.58	0.27	0.10	0.05	0.48	2.89

Monthly Results Suidkaap

	Flow mean	Flow std	Flow min	Flow max	bf mean	bf std	bf min	bf max	qf mean	qf std	qf min	qf max	BFI mean	BFI std	BFI min	BFI max	qf/bf
Month																	
January	3.80	2.95	0.55	11.51	1.85	1.32	0.34	5.03	1.96	2.09	0.17	9.58	0.54	0.18	0.17	0.89	1.06
February	4.19	5.57	0.35	27.98	2.08	1.88	0.26	8.24	**2.10**	3.88	0.05	19.74	0.60	0.17	0.28	0.95	1.01
March	3.73	3.88	0.53	19.28	**2.28**	2.08	0.26	9.55	1.44	1.96	0.06	9.73	0.64	0.17	0.31	0.96	0.63
April	2.75	3.02	0.46	15.14	1.99	1.89	0.28	10.21	0.75	1.35	0.01	6.60	0.76	0.16	0.42	0.99	0.38
May	1.74	1.58	0.25	8.63	1.52	1.39	0.19	7.40	0.22	0.22	0.03	1.23	0.86	0.06	0.74	0.99	0.14
June	1.34	0.98	0.30	5.04	1.12	0.92	0.18	4.73	0.22	0.13	0.06	0.62	0.80	0.09	0.59	0.97	0.20
July	1.20	0.82	0.28	4.18	0.96	0.74	0.18	3.81	0.23	0.12	0.06	0.49	0.78	0.07	0.64	0.91	0.24
August	1.06	0.74	0.21	3.39	0.79	0.62	0.04	3.04	**0.27**	0.24	0.01	1.34	0.73	0.16	0.16	0.98	0.34
September	1.03	0.73	0.16	2.91	0.71	0.50	0.12	2.32	0.32	0.36	0.03	1.61	0.71	0.15	0.33	0.93	0.45
October	1.15	0.76	0.19	3.04	**0.69**	0.49	0.09	2.19	0.46	0.38	0.10	1.55	0.60	0.13	0.34	0.87	0.66
November	1.98	1.84	0.27	8.90	0.89	0.68	0.10	2.96	1.09	1.25	0.17	6.05	0.48	0.15	0.27	0.84	1.23
December	3.37	2.92	0.44	11.58	1.55	1.33	0.25	5.49	1.82	1.72	0.19	7.64	0.48	0.14	0.15	0.72	1.18

Monthly Results Noordkaap

	Flow mean	Flow std	Flow min	Flow max	bf mean	bf std	bf min	bf max	qf mean	qf std	qf min	qf max	BFI mean	BFI std	BFI min	BFI max	qf/bf
Month																	
January	2.16	1.25	0.53	5.10	1.06	0.57	0.21	2.68	1.11	0.88	0.17	3.13	0.52	0.17	0.19	0.89	1.05
February	2.46	2.30	0.62	10.75	1.22	0.77	0.30	3.35	**1.24**	1.65	0.15	7.40	0.57	0.15	0.31	0.87	1.01
March	2.43	1.59	0.64	6.62	**1.47**	0.99	0.31	4.53	0.96	0.79	0.05	3.20	0.63	0.15	0.24	0.93	0.65
April	1.73	1.18	0.52	6.24	1.29	0.79	0.36	4.31	0.44	0.54	0.01	2.61	0.77	0.15	0.45	0.99	0.34
May	1.18	0.64	0.33	3.53	1.04	0.59	0.31	3.25	0.14	0.10	0.02	0.44	0.88	0.07	0.69	0.99	0.13
June	0.91	0.44	0.30	2.35	0.78	0.42	0.24	2.21	0.13	0.07	0.04	0.35	0.84	0.07	0.70	0.94	0.17
July	0.81	0.35	0.28	1.83	0.66	0.33	0.22	1.75	0.15	0.07	0.05	0.32	0.80	0.07	0.66	0.95	0.23
August	0.70	0.29	0.24	1.42	0.56	0.26	0.18	1.32	**0.15**	0.06	0.06	0.33	0.78	0.07	0.62	0.93	0.26
Septembe	0.66	0.34	0.20	1.64	0.46	0.21	0.14	1.00	0.19	0.20	0.04	0.90	0.73	0.13	0.41	0.94	0.41
October	0.72	0.34	0.19	1.43	**0.44**	0.20	0.14	0.89	0.28	0.21	0.04	1.08	0.63	0.11	0.24	0.79	0.65
Novembe	1.12	0.75	0.25	3.47	0.51	0.26	0.13	1.12	0.61	0.53	0.10	2.35	0.50	0.12	0.27	0.73	1.19
Decembe	1.79	1.13	0.27	4.34	0.80	0.48	0.17	2.03	0.99	0.70	0.10	2.67	0.47	0.12	0.24	0.72	1.24

Monthly Results Queens

	Flow mean	Flow std	Flow min	Flow max	bf mean	bf std	bf min	bf max	qf mean	qf std	qf min	qf max	BFI mean	BFI std	BFI min	BFI max	qf/bf
Month																	
January	3.00	3.05	0.23	12.79	2.24	2.08	0.20	10.07	0.76	1.19	0.01	6.20	0.80	0.11	0.47	0.97	0.34
February	2.78	3.38	0.07	15.34	2.28	2.61	0.06	11.49	**0.50**	0.83	0.01	3.85	0.85	0.07	0.70	0.99	0.22
March	2.44	2.10	0.22	7.82	**2.03**	1.70	0.16	6.55	0.42	0.45	0.02	1.94	0.83	0.08	0.67	0.96	0.21
April	1.53	1.59	0.20	7.20	1.37	1.38	0.17	6.27	0.16	0.23	0.01	0.99	0.89	0.06	0.73	0.99	0.12
May	0.78	0.72	0.09	2.72	0.74	0.69	0.08	2.63	0.04	0.04	0.00	0.18	0.93	0.04	0.84	0.98	0.05
June	0.53	0.48	0.04	1.72	0.50	0.47	0.04	1.68	0.03	0.02	0.00	0.08	0.94	0.03	0.86	0.98	0.05
July	0.44	0.40	0.03	1.40	0.41	0.37	0.02	1.36	0.03	0.04	0.00	0.19	0.92	0.05	0.77	0.97	0.07
August	0.34	0.31	0.03	1.26	0.32	0.30	0.03	1.19	**0.02**	0.02	0.00	0.09	0.91	0.07	0.69	0.98	0.08
Septembe	0.45	0.79	0.03	4.33	0.36	0.54	0.03	2.83	0.09	0.27	0.00	1.49	0.86	0.10	0.55	0.97	0.26
October	0.65	0.75	0.07	3.10	**0.51**	0.56	0.05	2.25	0.14	0.20	0.01	0.85	0.80	0.08	0.55	0.95	0.28
Novembe	1.44	1.58	0.03	6.69	1.11	1.17	0.02	4.91	0.33	0.43	0.01	1.78	0.79	0.08	0.65	0.96	0.30
Decembe	2.36	2.30	0.16	9.06	1.86	1.86	0.12	7.54	0.49	0.48	0.03	1.61	0.79	0.07	0.59	0.94	0.26

Annual Results Kaap

Hydro Year	Date	Rainfall	Flow	Baseflow bf	Quickflow qf	Ratio qf/bf	bfi_tr	Area (km^2) bfi_df	1640 Flow_mm	BF_mm	RC_flow	RC_BF
1979	30/09/1979	791.9	27.5	10.0	17.57	1.76	0.558	0.362	16.8	6.1	2%	1%
1980	30/09/1980	747.7	68.1	25.9	42.11	1.62	0.375	0.381	41.5	15.8	6%	2%
1981	30/09/1981	796.8	138.7	51.7	87.01	1.68	0.285	0.373	84.6	31.5	11%	4%
1982	30/09/1982	695	59.8	19.7	40.07	2.03	0.337	0.330	36.5	12.0	5%	2%
1983	30/09/1983	630.3	2.2	0.7	1.52	2.28	0.873	0.305	1.3	0.4	0%	0%
1984	30/09/1984	762	112.1	33.1	78.97	2.38	0.531	0.296	68.4	20.2	9%	3%
1985	30/09/1985	674	90.5	35.2	55.25	1.57	0.476	0.389	55.2	21.5	8%	3%
1986	30/09/1986	668.5	51.0	20.4	30.54	1.49	0.441	0.401	31.1	12.5	5%	2%
1987	30/09/1987	1229	29.4	8.6	20.79	2.42	0.506	0.292	17.9	5.2	1%	0%
1988	30/09/1988	769.8	112.9	42.2	70.71	1.68	0.391	0.374	68.8	25.7	9%	3%
1989	30/09/1989	919.9	72.9	28.1	44.82	1.60	0.435	0.385	44.5	17.1	5%	2%
1990	30/09/1990	668.1	92.3	35.5	56.77	1.60	0.390	0.385	56.3	21.7	8%	3%
1991	30/09/1991	1061.2	90.6	35.2	55.38	1.57	0.330	0.389	55.2	21.5	5%	2%
1992	30/09/1992	564.5	8.3	2.4	5.87	2.40	0.468	0.294	5.1	1.5	1%	0%
1993	30/09/1993	667	18.2	4.6	13.60	2.97	0.764	0.252	11.1	2.8	2%	0%
1994	30/09/1994	586	1.5	0.3	1.20	3.56	0.962	0.219	0.9	0.2	0%	0%
1995	30/09/1995	772.9	3.4	0.7	2.75	4.09		0.196	2.1	0.4	0%	0%
1996	30/09/1996	857.5	346.5	125.9	220.64	1.75	0.383	0.363	211.3	76.8	25%	9%
1997	30/09/1997	754	117.8	45.6	72.22	1.58	0.401	0.387	71.8	27.8	10%	4%
1998	30/09/1998	712.5	32.4	14.7	17.70	1.20	0.652	0.455	19.8	9.0	3%	1%
1999	30/09/1999	954	218.5	82.8	135.70	1.64	0.424	0.379	133.2	50.5	14%	5%
2000	30/09/2000	1265.7	596.4	237.9	358.50	1.51	0.457	0.399	363.6	145.0	29%	11%
2001	30/09/2001	691.5	208.7	84.6	124.12	1.47	0.423	0.405	127.3	51.6	18%	7%
2002	30/09/2002	693.7	111.8	46.8	65.03	1.39	0.445	0.418	68.2	28.5	10%	4%
2003	30/09/2003	590	34.3	11.3	23.07	2.05	0.644	0.328	20.9	6.9	4%	1%
2004	30/09/2004	679	55.0	19.7	35.37	1.80	0.576	0.357	33.6	12.0	5%	2%
2005	30/09/2005	950.5	22.7	7.0	15.67	2.23	0.685	0.309	13.8	4.3	1%	0%
2006	30/09/2006	750	138.4	55.7	82.75	1.49	0.572	0.402	84.4	33.9	11%	5%
2007	30/09/2007	652.5	33.3	12.2	21.12	1.73	0.574	0.366	20.3	7.4	3%	1%
2008	30/09/2008	745	48.7	18.1	30.59	1.69	0.512	0.372	29.7	11.1	4%	1%
2009	30/09/2009	815	66.4	26.0	40.47	1.56	0.355	0.391	40.5	15.8	5%	2%
2010	30/09/2010	933.5	191.4	77.6	113.84	1.47	0.325	0.405	116.7	47.3	13%	5%
2011	30/09/2011	821	209.7	85.5	124.18	1.45	0.391	0.408	127.9	52.1	16%	6%
2012	30/09/2012	702.9	82.7	32.7	49.96	1.53	0.662	0.396	50.4	19.9	7%	3%
	Min	564.5	1.5	0.3	1.2	1.20	0.285	0.196	0.9	0.2	0%	0%
	Max	1265.7	596.4	237.9	358.5	4.09	0.962	0.455	363.6	145.0	29%	11%
	Range	701.2	594.8	237.5	357.3	2.89	0.677	0.258	362.7	144.8	29%	11%
	Average	**781.6**	**102.8**	**39.4**	**63.4**	**1.89**	**0.503**	**0.358**	**62.7**	**24.0**	**7%**	**3%**
	SD	163.6	115.8	45.8	70.1	0.62	0.158	0.058	70.6	27.9	7%	3%
	Sum		3494.3	1338.4	2155.8	1.61						
	BFI and QF %			38%	62%							

Annual Results Suidkaap

				Baseflow	Quickflow	Ratio		Area (km^2)	262			
Hydro Year	Date	Rainfall	Flow	bf	qf	qf/bf	bfi_tr	bfi_df	Flow_mm	BF_mm	RC_flow	RC_BF
1979	30/09/1979	791.9	18.3	11.1	7.25	0.65	0.489	0.605	70.0	42.3	9%	5%
1980	30/09/1980	761.2	31.9	20.3	11.67	0.58	0.450	0.635	121.9	77.4	16%	10%
1981	30/09/1981	796.8	30.4	19.4	11.01	0.57	0.488	0.638	116.1	74.1	15%	9%
1982	30/09/1982	695	20.5	12.6	7.91	0.63	0.477	0.613	78.1	47.9	11%	7%
1983	30/09/1983	630.3	6.8	4.3	2.49	0.57	0.540	0.636	26.1	16.6	4%	3%
1984	30/09/1984	762	25.9	11.4	14.44	1.27	0.551	0.441	98.7	43.6	13%	6%
1985	30/09/1985	674	19.0	11.0	8.02	0.73	0.675	0.578	72.6	42.0	11%	6%
1986	30/09/1986	668.5	15.7	9.7	6.00	0.62	0.584	0.617	59.8	36.9	9%	6%
1987	30/09/1987	1229	11.1	5.5	5.57	1.01	0.670	0.498	42.3	21.1	3%	2%
1988	30/09/1988	769.8	25.7	15.2	10.51	0.69	0.744	0.591	98.1	58.0	13%	8%
1989	30/09/1989	919.9	20.8	11.8	8.95	0.76	0.713	0.569	79.2	45.1	9%	5%
1990	30/09/1990	668.1	27.2	16.4	10.84	0.66	0.783	0.601	103.8	62.4	16%	9%
1991	30/09/1991	1061.2	30.0	16.4	13.55	0.82	0.639	0.548	114.4	62.7	11%	6%
1992	30/09/1992	552	7.3	4.4	2.94	0.67	0.857	0.598	27.9	16.7	5%	3%
1993	30/09/1993	647	8.6	4.7	3.86	0.82	0.678	0.548	32.6	17.9	5%	3%
1994	30/09/1994	586	5.6	3.2	2.40	0.74	0.673	0.575	21.5	12.4	4%	2%
1995	30/09/1995	772.9	5.8	3.4	2.48	0.73	0.621	0.576	22.3	12.9	3%	2%
1996	30/09/1996	837.5	72.9	35.0	37.97	1.09	0.721	0.479	278.4	133.5	33%	16%
1997	30/09/1997	751	34.1	21.3	12.83	0.60	0.703	0.624	130.2	81.2	17%	11%
1998	30/09/1998	712.5	17.0	12.3	4.76	0.39	0.715	0.721	65.1	46.9	9%	7%
1999	30/09/1999	954	44.8	25.7	19.14	0.75	0.683	0.573	171.0	97.9	18%	10%
2000	30/09/2000	1265.7	93.7	56.7	36.99	0.65	0.704	0.605	357.4	216.3	28%	17%
2001	30/09/2001	691.5	55.4	36.8	18.54	0.50	0.652	0.665	211.3	140.5	31%	20%
2002	30/09/2002	693.7	33.9	23.1	10.80	0.47	0.554	0.682	129.5	88.3	19%	13%
2003	30/09/2003	590	20.8	13.4	7.39	0.55	0.558	0.644	79.3	51.1	13%	9%
2004	30/09/2004	679	19.1	11.0	8.09	0.73	0.805	0.577	72.9	42.0	11%	6%
2005	30/09/2005	950.5	9.7	5.5	4.19	0.76	0.896	0.567	37.0	21.0	4%	2%
2006	30/09/2006	750	31.8	19.7	12.09	0.61	0.755	0.620	121.4	75.3	16%	10%
2007	30/09/2007	652.5	14.7	8.8	5.91	0.67	0.849	0.599	56.2	33.7	9%	5%
2008	30/09/2008	745	19.0	12.0	7.01	0.58	0.721	0.632	72.7	46.0	10%	6%
2009	30/09/2009	815	22.6	14.9	7.72	0.52	0.551	0.659	86.4	56.9	11%	7%
2010	30/09/2010	933.5	52.5	32.2	20.25	0.63	0.658	0.614	200.4	123.1	21%	13%
2011	30/09/2011	821	55.6	35.8	19.77	0.55	0.641	0.644	212.1	136.6	26%	17%
2012	30/09/2012	702.9	20.7	13.8	6.90	0.50	0.422	0.667	79.1	52.8	11%	8%
	Min	552	5.6	3.2	2.4	0.39	0.422	0.441	21.5	12.4	3%	2%
	Max	1265.7	93.7	56.7	38.0	1.27	0.896	0.721	357.4	216.3	33%	20%
	Range	713.7	88.0	53.4	35.6	0.88	0.474	0.279	335.9	203.9	30%	19%
	Average	**780.3**	**27.3**	**16.4**	**10.9**	**0.68**	**0.653**	**0.601**	**104.3**	**62.7**	**13%**	**8%**
	SD	164.3	19.6	11.7	8.4	0.17	0.118	0.056	74.9	44.6	8%	5%
	Sum		929.0	558.8	370.2	0.66						
	BFI and QF %			60%	40%							

Annual Results Noordkaap

				Baseflow	Quickflow	Ratio		Area (km^2)	126			
Hydro Year	Date	Rainfall	Flow	bf	qf	qf/bf	bfi_tr	bfi_df	Flow_mm	BF_mm	RC_flow	RC_BF
1979	30/09/1979	791.9	9.3	5.9	3.35	0.57	0.638	0.638	73.5	46.9	9%	6%
1980	30/09/1980	747.7	15.9	10.0	5.83	0.58	0.505	0.632	125.8	79.5	17%	11%
1981	30/09/1981	796.8	19.4	11.9	7.47	0.63	0.507	0.615	154.1	94.8	19%	12%
1982	30/09/1982	695	16.5	10.4	6.10	0.59	0.560	0.629	130.6	82.2	19%	12%
1983	30/09/1983	630.3	7.5	5.1	2.40	0.47	0.605	0.681	59.7	40.7	9%	6%
1984	30/09/1984	762	16.7	9.4	7.28	0.78	0.577	0.563	132.3	74.5	17%	10%
1985	30/09/1985	674	17.4	10.9	6.47	0.59	0.626	0.628	138.1	86.8	20%	13%
1986	30/09/1986	668.5	19.0	12.0	6.95	0.58	0.591	0.634	150.6	95.5	23%	14%
1987	30/09/1987	1229	13.0	7.3	5.65	0.77	0.688	0.565	103.0	58.2	8%	5%
1988	30/09/1988	769.8	23.7	14.4	9.30	0.65	0.610	0.607	188.0	114.2	24%	15%
1989	30/09/1989	919.9	18.1	11.0	7.10	0.64	0.636	0.608	143.9	87.6	16%	10%
1990	30/09/1990	668.1	18.9	11.9	7.07	0.59	0.678	0.627	150.4	94.3	23%	14%
1991	30/09/1991	1061.2	21.7	12.8	8.84	0.69	0.735	0.592	171.9	101.8	16%	10%
1992	30/09/1992	552	6.5	4.3	2.14	0.49	0.863	0.669	51.4	34.4	9%	6%
1993	30/09/1993	647	7.4	4.4	2.99	0.68	0.794	0.594	58.3	34.6	9%	5%
1994	30/09/1994	586	5.4	3.3	2.11	0.64	0.787	0.609	42.9	26.1	7%	4%
1995	30/09/1995	772.9	5.0	2.9	2.04	0.70	0.875	0.590	39.5	23.3	5%	3%
1996	30/09/1996	837.5	31.3	17.5	13.83	0.79	0.656	0.559	248.8	139.0	30%	17%
1997	30/09/1997	751	19.1	11.7	7.46	0.64	0.724	0.610	151.9	92.7	20%	12%
1998	30/09/1998	712.5	12.7	8.7	4.05	0.47	0.732	0.682	101.2	69.0	14%	10%
1999	30/09/1999	954	24.7	15.0	9.73	0.65	0.681	0.606	196.2	119.0	21%	12%
2000	30/09/2000	1265.7	40.2	24.5	15.65	0.64	0.603	0.611	319.0	194.8	25%	15%
2001	30/09/2001	691.5	21.9	14.5	7.42	0.51	0.728	0.662	174.2	115.3	25%	17%
2002	30/09/2002	693.7	16.6	10.6	5.99	0.57	0.586	0.638	131.4	83.9	19%	12%
2003	30/09/2003	590	10.5	6.3	4.24	0.67	0.680	0.598	83.7	50.1	14%	8%
2004	30/09/2004	679	9.7	5.5	4.11	0.74	0.739	0.574	76.6	44.0	11%	6%
2005	30/09/2005	950.5	8.5	5.1	3.41	0.67	0.847	0.599	67.5	40.5	7%	4%
2006	30/09/2006	750	20.0	12.5	7.54	0.61	0.740	0.623	158.7	98.9	21%	13%
2007	30/09/2007	652.5	10.8	6.4	4.41	0.69	0.862	0.592	85.7	50.7	13%	8%
2008	30/09/2008	745	12.3	7.8	4.54	0.58	0.864	0.632	97.9	61.9	13%	8%
2009	30/09/2009	815	16.5	10.3	6.15	0.60	0.850	0.627	130.7	81.9	16%	10%
2010	30/09/2010	933.5	24.2	15.3	8.83	0.58	0.931	0.635	191.7	121.7	21%	13%
2011	30/09/2011	821	30.5	19.8	10.75	0.54	0.768	0.648	242.2	156.9	30%	19%
2012	30/09/2012	702.9	15.9	10.3	5.68	0.55	0.921	0.643	126.5	81.4	18%	12%
	Min	552	5.0	2.9	2.0	0.47	0.505	0.559	39.5	23.3	5%	3%
	Max	1265.7	40.2	24.5	15.7	0.79	0.931	0.682	319.0	194.8	30%	19%
	Range	713.7	35.2	21.6	13.6	0.32	0.426	0.123	279.5	171.5	25%	16%
	Average	**779.9**	**16.7**	**10.3**	**6.4**	**0.62**	**0.711**	**0.618**	**132.3**	**81.7**	**17%**	**10%**
	SD	164.4	7.9	4.8	3.1	0.08	0.118	0.031	62.6	38.3	6%	4%
	Sum		566.8	349.9	216.9	0.62						
	BFI and QF %			62%	38%							

Annual Results Queens

				Baseflow	Quickflow	Ratio		Area (km^2)	180			
Hydro Year	Date	Rainfall	Flow	bf	qf	qf/bf	bfi_tr	bfi_df	Flow_mm	BF_mm	RC_flow	RC_BF
1979	30/09/1979	791.9	3.0	2.4	0.61	0.25	0.632	0.800	16.9	13.5	2%	2%
1980	30/09/1980	747.7	12.1	9.5	2.56	0.27	0.581	0.788	67.1	52.9	9%	7%
1981	30/09/1981	796.8	21.1	16.6	4.50	0.27	0.598	0.786	117.1	92.1	15%	12%
1982	30/09/1982	695	9.9	8.1	1.81	0.22	0.592	0.817	55.0	45.0	8%	6%
1983	30/09/1983	630.3	1.8	1.4	0.37	0.27	0.546	0.788	9.8	7.7	2%	1%
1984	30/09/1984	762	24.8	16.7	8.12	0.49	0.575	0.673	138.0	92.9	18%	12%
1985	30/09/1985	674	21.2	17.2	3.99	0.23	0.592	0.812	117.7	95.5	17%	14%
1986	30/09/1986	668.5	8.5	7.1	1.45	0.21	0.544	0.829	47.3	39.2	7%	6%
1987	30/09/1987	1229	5.6	4.1	1.55	0.38	0.493	0.723	31.2	22.6	3%	2%
1988	30/09/1988	769.8	19.6	15.9	3.71	0.23	0.805	0.810	108.8	88.2	14%	11%
1989	30/09/1989	919.9	15.2	12.7	2.53	0.20	0.722	0.834	84.6	70.6	9%	8%
1990	30/09/1990	668.1	13.6	11.5	2.08	0.18	0.695	0.847	75.5	64.0	11%	10%
1991	30/09/1991	1061.2	20.8	17.0	3.83	0.23	0.731	0.816	115.5	94.2	11%	9%
1992	30/09/1992	552	2.7	2.1	0.55	0.26	0.836	0.794	14.9	11.9	3%	2%
1993	30/09/1993	647	4.0	3.0	1.07	0.36	0.747	0.736	22.4	16.5	3%	3%
1994	30/09/1994	586	4.1	3.3	0.76	0.23	0.819	0.812	22.5	18.3	4%	3%
1995	30/09/1995	772.9	3.6	2.9	0.74	0.26	0.630	0.795	20.2	16.0	3%	2%
1996	30/09/1996	837.5	39.8	31.8	7.97	0.25	0.715	0.800	221.2	176.9	26%	21%
1997	30/09/1997	751	22.2	18.6	3.63	0.20	0.757	0.836	123.3	103.1	16%	14%
1998	30/09/1998	712.5	7.9	7.0	0.94	0.13	0.687	0.882	44.1	38.8	6%	5%
1999	30/09/1999	954	18.3	15.3	2.95	0.19	0.783	0.838	101.5	85.1	11%	9%
2000	30/09/2000	1265.7	46.4	38.8	7.57	0.19	0.770	0.837	257.8	215.7	20%	17%
2001	30/09/2001	691.5	24.1	20.8	3.32	0.16	0.611	0.862	134.0	115.5	19%	17%
2002	30/09/2002	693.7	21.1	18.1	2.96	0.16	0.551	0.860	117.1	100.7	17%	15%
2003	30/09/2003	590	10.2	8.3	1.89	0.23	0.428	0.814	56.5	46.0	10%	8%
2004	30/09/2004	679	9.9	8.2	1.72	0.21	0.640	0.826	55.1	45.6	8%	7%
2005	30/09/2005	950.5	5.9	5.0	0.81	0.16	0.846	0.861	32.5	28.0	3%	3%
2006	30/09/2006	750	24.4	19.8	4.66	0.24	0.722	0.809	135.7	109.8	18%	15%
2007	30/09/2007	652.5	4.9	4.0	0.90	0.22	0.777	0.816	27.1	22.1	4%	3%
2008	30/09/2008	750	16.4	13.5	2.96	0.22	0.631	0.820	91.2	74.7	12%	10%
2009	30/09/2009	815	18.9	15.3	3.61	0.24	0.660	0.809	105.1	85.1	13%	10%
2010	30/09/2010	933.5	37.9	32.3	5.64	0.17	0.859	0.851	210.7	179.4	23%	19%
2011	30/09/2011	821	49.5	42.3	7.20	0.17	0.619	0.854	274.8	234.8	33%	29%
2012	30/09/2012	702.9	19.8	16.1	3.72	0.23	0.759	0.813	110.2	89.6	16%	13%
Min		552	1.8	1.4	0.4	0.13	0.428	0.673	9.8	7.7	2%	1%
Max		1265.7	49.5	42.3	8.1	0.49	0.859	0.882	274.8	234.8	33%	29%
Range		713.7	47.7	40.9	7.7	0.35	0.430	0.209	265.0	227.0	32%	27%
Average		**780.1**	**16.7**	**13.7**	**3.0**	**0.23**	**0.675**	**0.813**	**93.0**	**76.2**	**12%**	**10%**
SD		164.3	12.4	10.3	2.2	0.07	0.108	0.041	68.7	57.4	8%	6%
Sum			569.3	466.6	102.7	0.22						
BFI and QF %				82%	18%							

A3 (Chapter 7) – Physiographic and hydroclimatic characteristics of Kaap catchment and tributaries.

Streamgauge		Noordkaap X2H010	Queens X2H008	Suidkaap X2H031	Kaap Total X2H022
Sub-basin area (km^2)		126	180	262	1640
HAND zones	Wetland	8%	7%	7%	8%
	Plateau	51%	32%	61%	39%
	Hillslope	42%	61%	32%	53%
Soil texture	Clay	19%	5%	5%	4%
	Sandy clay	4%	7%	9%	4%
	Clay loam	42%	42%	25%	39%
	Sandy clay loam	34%	46%	60%	53%
	Sandy loam	0%	0%	1%	0%
Geology	Granite	97%	58%	98%	52%
	Lava	0%	28%	1%	16%
	Arenite	0%	2%	0%	9%
	Ultramafic rocks	0%	4%	0%	2%
	Quartzite	3%	0%	0%	0%
	Gneiss	0%	0%	1%	6%
	Lutaceous arenite	0%	7%	0%	14%
LULC	Forest/Woodland	14%	12%	9%	20%
	Bush/Shrub	11%	9%	17%	32%
	Grassland	7%	18%	10%	14%
	Plantations	62%	60%	52%	23%
	Water	0%	0%	0%	0%
	Wetlands	0%	1%	1%	1%
	Bare	0%	0%	0%	0%
	Agriculture: Rainfed, Planted pasture, Fallow	2%	0%	5%	3%
	Agriculture: Irrigated	3%	0%	5%	6%
	Urban and Mines	0%	0%	0%	2%
Mean Annual Runoff observed (mm/y)		149	99	120	66
Mean Annual Runoff naturalized (mm/y) [a]		216	146	210	116
Mean annual Precipitation (mm/y)		1101	1016	905	900
Mean annual potential evaporation (mm/y)		1425	1369	1451	1435

A4 (Chapter 7) – Supplementary material

S1. Model Input

Precipitation

The station data from several sources was assessed for completeness and consistency. SAWS, Lynch database, SASRI, ICMA and DWA data were checked. SAWS was found more consistent and used. SASRI data was used to complement - especially where gaps existed on SAWS records. Linear regression was used to infill stations with the most correlated neighbouring station data.

We also looked at remote sensing data for rainfall. CHIRPS, CMORPH and TRMM daily data was obtained. The data was aggregated to monthly and annual totals for comparison with station data and Mean Annual Precipitation map. Due to coarse resolution of the CMORPH and TRMM only the CHIRPS dataset was for model input.

Evaporation

We looked at evaporation data from the ground weather stations of SASRI. We also looked at the remote sensing products ALEXI, CMRSET and SSEBop. These products had different temporal and spatial resolutions. Table 1 shows an overview of remote sensing products analysed.

Table 8. Overview of Remote sensing products used

	Product	Spatial resolution	Temporal resolution	Period covered	Source/Literature
Precipitation	CHIRPS	0.05x0.05 degrees	Daily	2000/01/01 – 2013/12/31	Funk *et al.* (2015)
	CMORPH	0.25x0.25 degrees	Daily	2000/01/01 – 2013/12/32	
	TRMM	0.25x0.25 degrees	Daily	2000/01/01 – 2013/12/33	
Evaporation	ALEXI	0.05x0.05 degrees	Weekly	2003/01/01 – 2013/12/24	Anderson *et al.* (1997);Hain *et al.* (2009)
	CMRSET	0.05x0.05 degrees	Monthly	2000/01/01 – 2012/12/01	Guerschman *et al.* (2009)
	SSEBop	0.0083x0.0083 degrees (90x90m)	Monthly	2003/01/01 – 2013/12/01	Senay *et al.* (2013); Chen *et al.* (2016)

Soil data

Different sources of soil data are available for modelling in South Africa, and Southern Africa. Paterson *et al.* (2015) provides a comprehensive review of history and development of soil information in South Africa. An overview of different soil data and soil derived parameter sources are listed on Table 2.

Table 9. Soil data sources and products available

Database	Source	Scale	Grid/polygon	Coverage	Reference
Land types of South Africa	ARC-ISCW, AGIS	1:250 000	Polygons	National (SA)	Group and Macvicar (1991)
Harmonized World Soil Database	FAO	1:5 000 000	30 arc-second	World	Nachtergaele *et al.* (2008)
Atlas ACRU (derived from Land types)	Atlas	1:250 000	Polygons	National (SA)	Schulze *et al.* (2007)
SOTERSAF	ISRIC	1:1 000 000	Polygons	Southern Africa	van Engelen and Dijkshoorn (2013)
Soil Grids 1km	ISRIC	1:1 000 000	1km grid	World	Hengl *et al.* (2014)
Soil Grids 250m	AfSIS/ISRIC	1:250 000	250m grid	Africa	Hengl *et al.* (2015)

ARC-ISCW – Agricultural Research Council - Institute for Soil, Climate and Water
AGIS – Agricultural Geo-referenced Information System
FAO – Food and Agriculture Organization
ISRIC – World Soil Information

Land type survey of South Africa (Group and Macvicar, 1991) is the most commonly used in South Africa. It divides South Africa into a number of unique mapping units, or land types, each with a unique combination of soil pattern, macroclimate and terrain form. The extensive survey was conducted at 1:250,000 scale. However, the derivation of hydrological parameters from the database is not straightforward, and different hydrological models have used different approaches.

The South African atlas of climatology and agro-hydrology (Schulze *et al.*, 2007) database contains soil data derived from the land types of South Africa (Group and Macvicar, 1991). Schulze (1985) and Schulze *et al.* (2007) derived relevant hydrological parameters from the soil data using AUTOSOILS decision support tool (Pike and Schulze, 1995).

The PITMAN model made a simplification of the land types using their lithology and soil texture, and has also derived typical hydrological parameters from the same database.

The soil and terrain database for South Africa (SOTERSAF) was also derived from the land types using SOTER methodology (van Engelen and Dijkshoorn, 2013), in order to harmonize it to the rest of Southern Africa and with world standards. This database was compiled by ISRIC - World Soil Information under the framework of the Land Degradation Assessment in Drylands (LADA, GLADA) program. The initial dataset was compiled by the Institute of Soil, Climate and Water (ISCW), Pretoria, at scale 1:1,000,000, which means some details and information, was aggregated.

The Soil Grids initiative (Hengl *et al.*, 2014), also lead by ISRIC, aims at further standardizing soil data and soil derived parameters, for application in agricultural and hydrological models and products. Initially, the Soil Grids 1km was developed, and now more refined Soil Grids 250m (Hengl *et al.*, 2015; Hengl *et al.*, 2017) is also available.

In this research we tested the different sources of soil data in a hydrological model, to see whether recent developments in the provision of soil data, particularly the Soil Grids 250m dataset (Figure 2 and Table 1), improve hydrological simulations. This is particularly relevant for trans-boundary river basins, such as the Incomati River basin, given that the available soil data is derived from three different countries databases, which are not harmonized. Therefore, Soil Grids could provide a consistent input data set to model the entire trans-boundary basin.

S2. Results of the selected four STREAM model runs

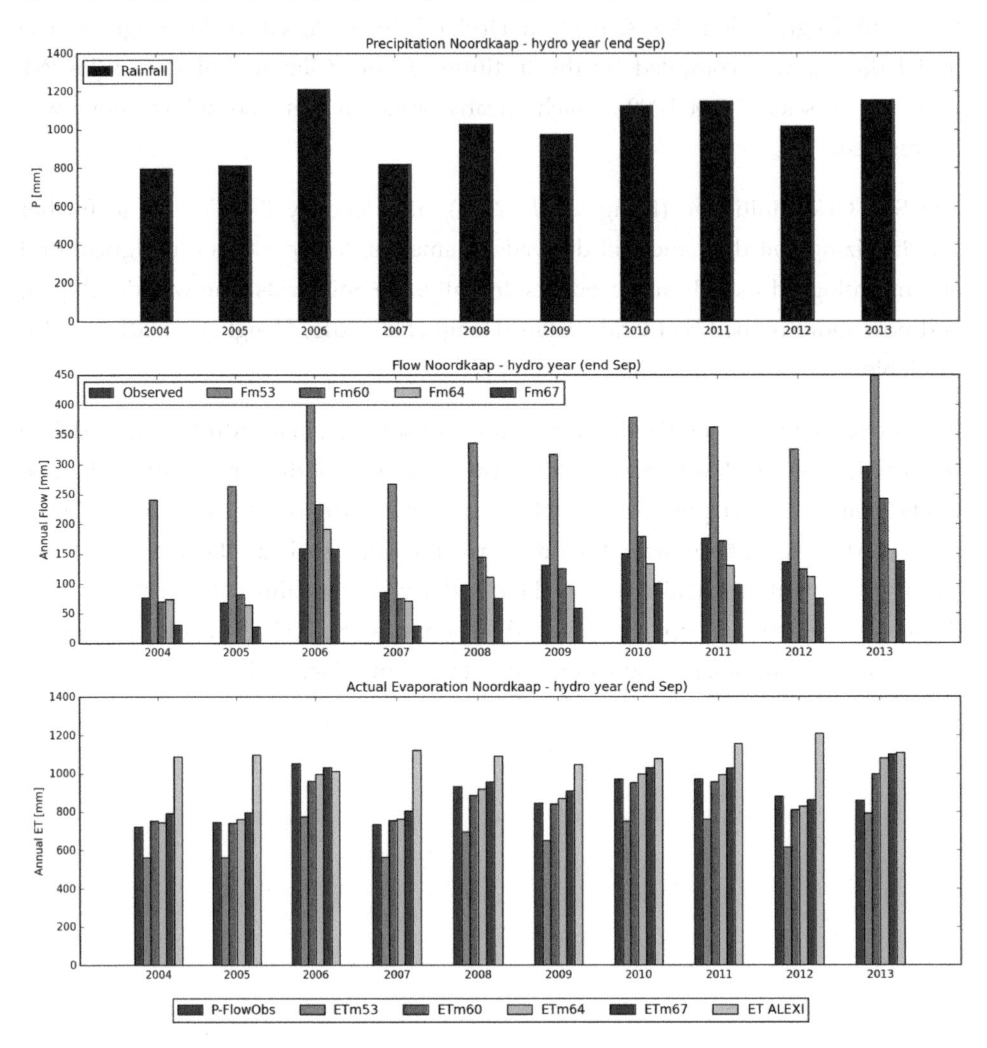

Figure 9. Annual Water Balance of the Noordkaap catchment. The subscripts of flow and evaporation refer to model simulations presented in the main text.

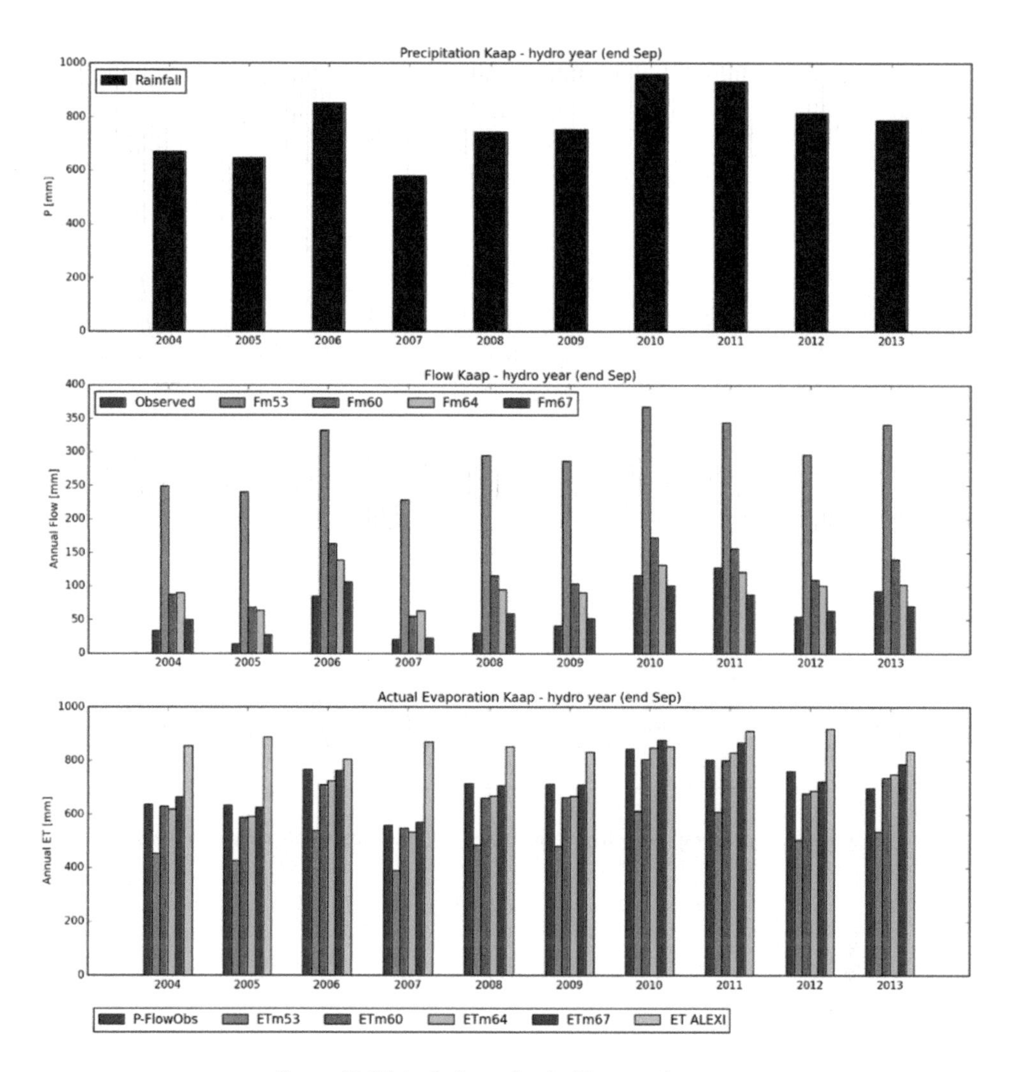

Figure 10. Water balance for the Kaap catchment.

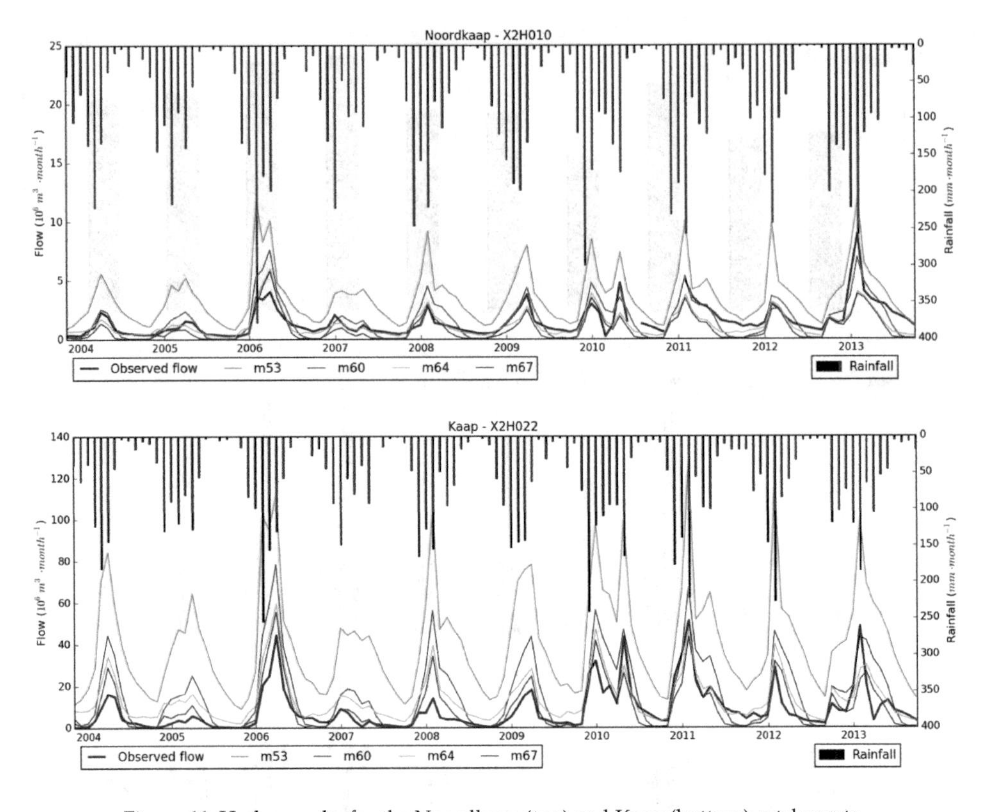

Figure 11. Hydrographs for the Noordkaap (top) and Kaap (bottom) catchments.

S3. Comparison of STREAM and HBV results

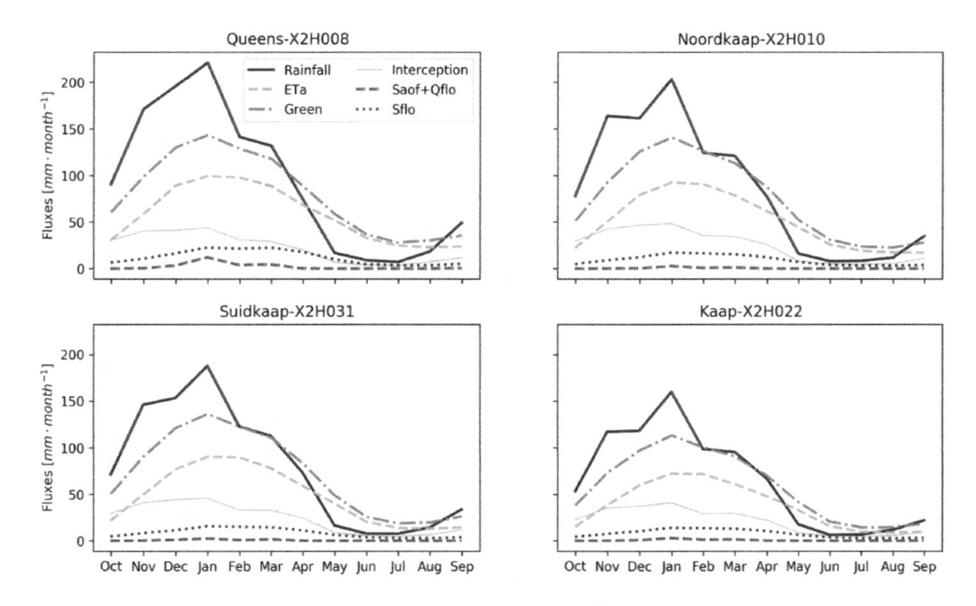

Figure 12. Mean monthly water balance and flow components for the four catchments, using results of run 64. Eta is actual evaporation, Green is the total evaporation (including interception), Saof is saturated overland flow, Qflo is quickflow component and Sflo is the slow flow (or baseflow) component

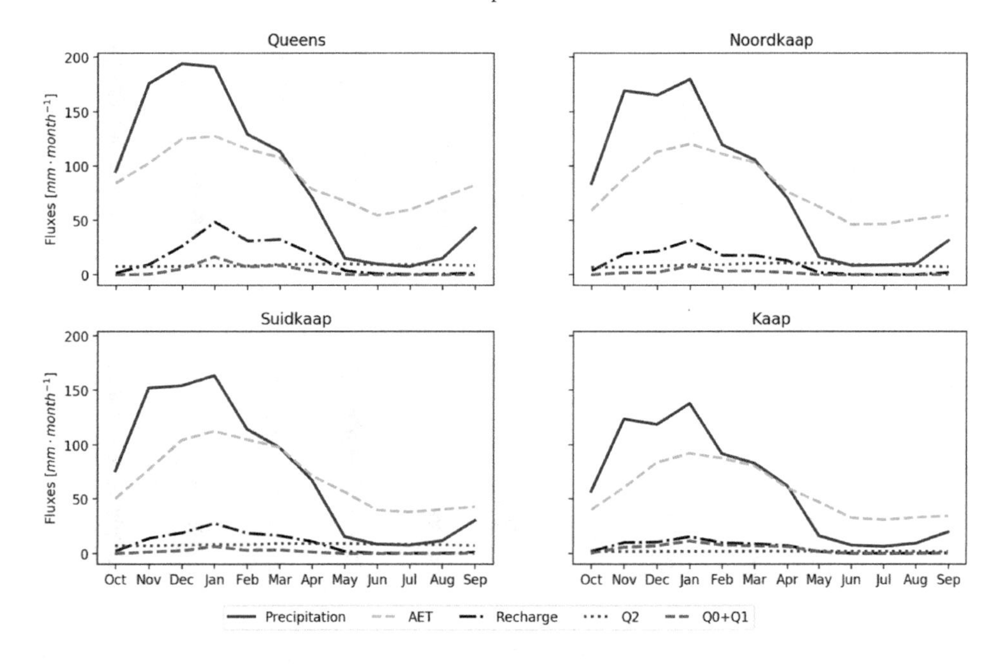

Figure 13. Average water balance of HBV model results. AET stands for actual evaporation and Q0, Q1 and Q2 are the flow components, fastest to slowest.

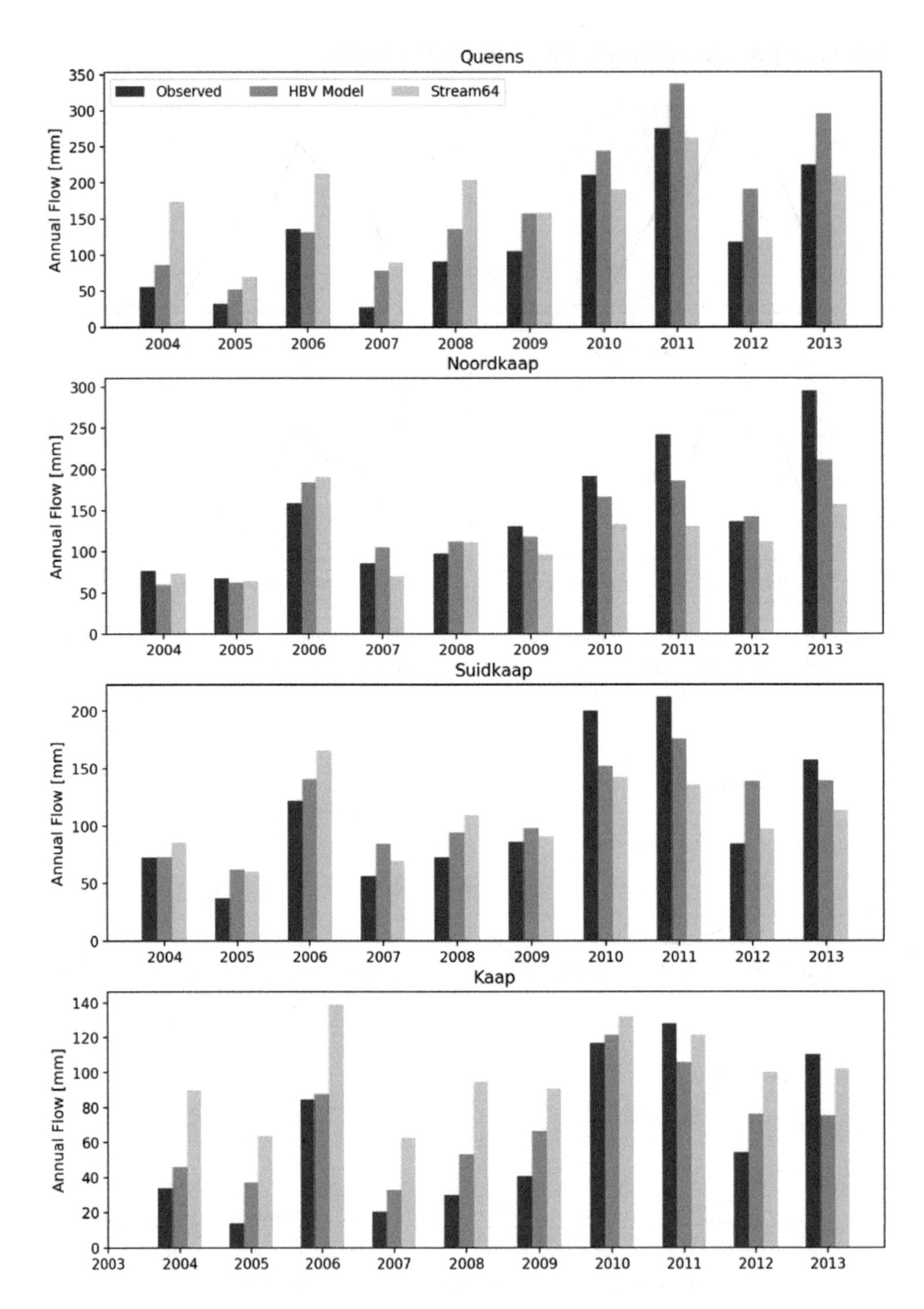

Figure 14. Comparison of annual flows observed and simulated by HBV and Stream (run64) models

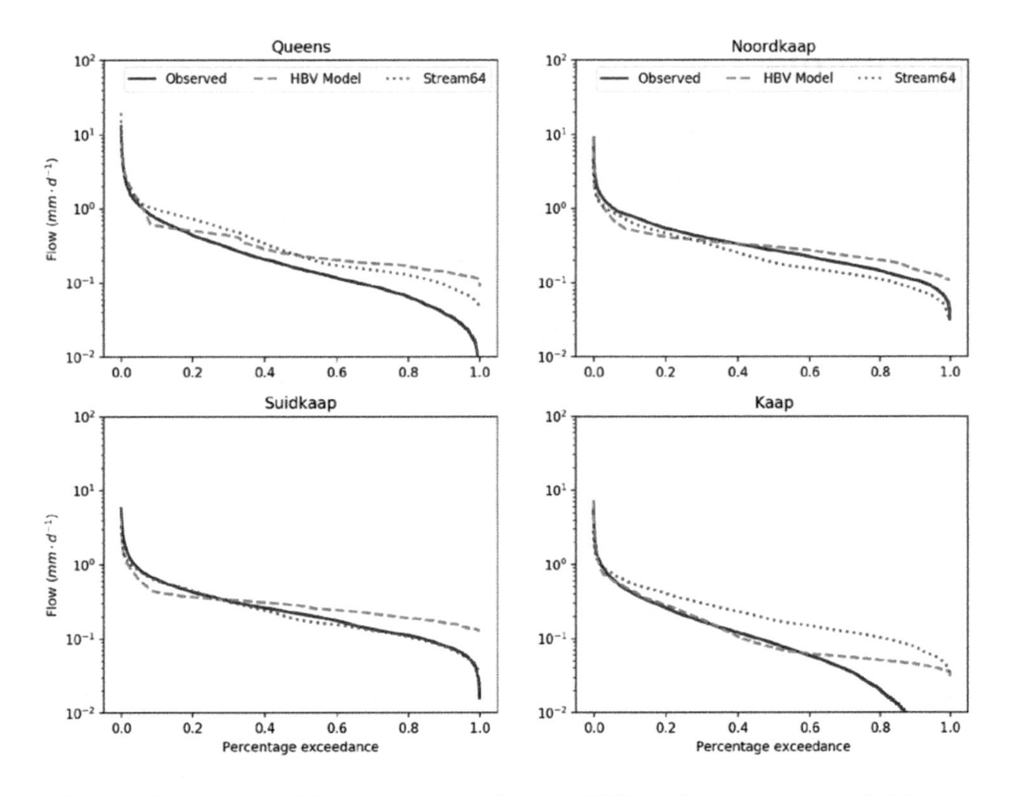

Figure 15. Comparison of FDCs from observed flow and HBV and Stream run 64 modelled flows

Acronyms

α	Recession constant
ACRU	Agrohydrological Model
AGIS	Agricultural Geo-referenced Information System
ALEXI	Atmosphere–Land Exchange Inverse Model
ANOVA	One-way ANalysis Of Variance
API	Antecedent Precipitation Index
ARA-Sul	Administração Regional de Águas - South Regional Water Administration (Mozambique)
ARC-ISCW	Agricultural Research Council - Institute for Soil, Climate and Water
BFI	Baseflow index
BFI_{max}	Maximum value of the baseflow index (the long term ratio of baseflow to river discharge)
CD	Coefficient of dispersion
CHIRPS	Climate Hazards Group InfraRed Precipitation with Station data
CMRSET	Commonwealth Scientific and Industrial Research Organisation (CSIRO) Moderate Resolution Imaging Spectroradiometer (MODIS) Reflectance Scaling EvapoTranspiration (CMRSET)
CWP	Crop Water Productivity
DEM	Digital Elevation Model
DF	Digital filter
DNA	Direcção Nacional de Águas - National Water Directory (Mozambique)
DOY	Day of the year
DRP	Dominating runoff processes
DSS	Decision Support Systems
DWAF/DWA	Department of Water Affairs and Forestry (renamed Department of Water Affairs, and currently Department of Water and Sanitation - DWS)
EC	Electrical Conductivity
EMMA	End Member Mixing Analysis
ESKOM	South African electricity public utility
EWP	Economic Water Productivity
FAO	Food and Agriculture Organization
GIS	Geographical Information System

GPS	Global Positioning System
GMWL	Global Meteoric Water Line
HAND	Height Above Nearest Drainage
HBV	Hydrologiska Byråns Vattenbalansavdelning (model)
IAAP	Implementation Activity and Action Plan
IAEA	International Atomic Energy Agency
IAHS	International Association of Hydrological Sciences
IDW	Inverse Distance Weighing method
IUCMA/ICMA	Inkomati-Usuthu Catchment Management Agency (former Inkomati Catchment Management Agency)
IIMA	Interim Inco-Maputo Agreement
ISRIC	International Soil Reference and Information Centre
ISP	Internal Strategic Perspective
IWAAS	Inkomati Water Availability Assessment Study
JIBS	Joint Incomati Basin Study
KGE	Kling-Gupta efficiency
KNP	Kruger National Park
KOBWA	Komati Water Basin Authority
LMWL	Local Mean Water Line
LogNSE	Logarithmic Nash-Sutcliffe efficiency
LULC	Land Use and Land Cover
MAE	Mean absolute error
MAR	Mean Annual Runoff
NSE	Nash-Sutcliffe Efficiency
ORP	Oxidation and Redox Potential
PBias	Percentage Bias
PCA	Principal Component Analysis
PET	Potential evapotranspiration
PRIMA	Progressive Realisation of the Inco-Maputo Agreement
PUB	Prediction of Ungauged Basins
PWTW	Potable Water Treatment Works
RISKOMAN	Risk-based Operational Water Management for the Incomati River Basin
RMSE	Root Mean Square Error
SANBI	South Africa National Botanical Institute
SANParks	South African National Parks

SASRI	South African Sugarcane Research Institute
SAWS	South African Weather Service
SDG	Sustainable Development Goal
SEBAL	Surface Energy Balance Algorithm for Land
SEI	Stockholm Environment Institute
SENSE	Research School for Socio-Economic and Natural Sciences of the Environment
SFRA	Streamflow Reduction Activity
SSEBop	Operational Simplified Surface Energy Balance (SSEBop)
STRM	Shuttle Radar Topography Mission
STREAM	Spatial Tools for River basin Environmental Analysis and Management
SWAT	Soil and Water Assessment Tool
TAC	Tracer Aided Catchment Model
TIA	Tripartite Interim Agreement between Mozambique, South Africa and Swaziland
TPTC	Tripartite Permanent Technical Committee
UKZN	University of KwaZulu-Natal
UPaRF	UNESCO-IHE Partnership Research Fund
VSMOW	Vienna Standard Mean Ocean Water
WAFLEX	Spreadsheet-based water resources model
WAS	Water Accounting System
WATPLAN	Spatial earth observation monitoring for planning and water allocation in the international Incomati Basin
WEAP	Water Evaluation and Planning Model
WMA	Water Management Area
WRC	Water Research Commission of South Africa
WReMP	Water Resources Modelling Platform
WRSM	Water Resource Management System Model
WRYM	Water Resources Yield Model
WWTW	Waste Water Treatment Work

Biography

Aline Maraci Lopes Saraiva Okello, graduated with distinction from the MSc Programme in Water Science and Engineering, specialisation Hydrology and Water Resources, from IHE Delft Institute for Water Education (former UNESCO-IHE), Delft, The Netherlands, in April 2010. Her MSc research topic was "Experimental Investigation of Water Fluxes in Irrigated Sugarcane using Environmental Isotopes. A case Study of Mhlume Plantations, Incomati Catchment, Swaziland". The research was carried out under the RISKOMAN project (Risk-based operational water management on the Incomati River Basin). This project was financed by the Water Research Commission of South Africa (WRC) and IHE Delft (through funding obtained from DGIS, Netherlands). Aline was selected to continue as a PhD researcher for the RISKOMAN project.

Aline holds a BSc Honours degree (Licenciatura) in Civil and Transport Engineering (with distinction: 90%), from ISUTC, Maputo, Mozambique, with thesis: Drainage Systems applied to a restricted area of Maputo city. She has been a teacher, researcher and consultant over the past 10 years. She is member of IAHS (International Association of Hydrological Sciences) and WISA (Water Institute of Southern Africa and Mozambican Council of Engineers.

She received the L'Oreal-UNESCO For Women in Science Sub-Saharan Africa Fellowship Award in 2013 and Faculty for the Future Fellowship in 2014 and 2015. She was shortlisted for the Africa Prize for Engineering Innovation 2016/2017 for the development of the mobile application HarvestRainWater. Aline was also selected for the first cohort of the UNLEASH Global Innovation Lab for the SDGs in Denmark, August 2017. In general, she is very interested in research and development, and women & youth empowerment.

List of publications

Journal papers

Saraiva Okello AML, Masih I, Uhlenbrook S, Jewitt GPW, Van der Zaag P. 2018. Improved Process Representation in the Simulation of the Hydrology of a Meso-Scale Semi-Arid Catchment. Water, 10: 1549. DOI: https://doi.org/10.3390/w10111549.

Saraiva Okello AML, Uhlenbrook S, Jewitt GPW, Masih I, Riddell ES, Van der Zaag P. 2018. Hydrograph separation using tracers and digital filters to quantify runoff components in a semi-arid mesoscale catchment. Hydrological Processes, 32: 1334 - 1350. DOI: doi:10.1002/hyp.11491.

Saraiva Okello AML, Masih I, Uhlenbrook S, Jewitt GPW, van der Zaag P, Riddell E. 2015. Drivers of spatial and temporal variability of streamflow in the Incomati River basin. Hydrol. Earth Syst. Sci., 19: 657-673. DOI: 10.5194/hess-19-657-2015.

Camacho Suarez VV, **Saraiva Okello AML**, Wenninger JW, Uhlenbrook S. 2015. Understanding runoff processes in a semi-arid environment through isotope and hydrochemical hydrograph separations. Hydrol. Earth Syst. Sci., 19: 4183-4199. DOI: 10.5194/hess-19-4183-2015.

van Eekelen MW, Bastiaanssen WGM, Jarmain C, Jackson B, Ferreira F, van der Zaag P, **Saraiva Okello A**, Bosch J, Dye P, Bastidas-Obando E, Dost RJJ, Luxemburg WMJ. 2015. A novel approach to estimate direct and indirect water withdrawals from satellite measurements: A case study from the Incomati basin. Agriculture, Ecosystems & Environment, 200: 126-142. DOI: http://dx.doi.org/10.1016/j.agee.2014.10.023.

Other publications

E.S. Riddell, G.P.W. Jewitt, T.K. Chetty, **A.M.L. Saraiva Okello**, B. Jackson, A. Lamba, S. Gokool, P. Naidoo, T. Vather, S.Thornton-Dibb, Final Report. A Management Tool for the Inkomati Basin with focus on Improved Hydrological Understanding for Risk-based Operational Water Management. Deliverable 9, Project K5/1935, March 2014.

E.S. Riddell, **A.M.L. Saraiva Okello**, T.K. Chetty, S.Thornton-Dibb, B. Jackson, G.P.W. Jewitt. Annual Report 3. A Management Tool for the Inkomati Basin with focus on Improved Hydrological Understanding for Risk-based Operational Water Management. Deliverable 7, Project K5/1935, March 2013.

E.S. Riddell, G.P.W. Jewitt, **A.M.L. Saraiva Okello**, T.K. Chetty, B. Jackson. Annual Report 2. A Management Tool for the Inkomati Basin with focus on Improved Hydrological Understanding for Risk-based Operational Water Management. Deliverable 4, Project K5/1935, March 2012.

E.S. Riddell, T.K. Chetty, **A.M.L. Saraiva Okello**, B. Jackson, G.P.W. Jewitt. Report on new sources of catchment information. A Management Tool for the Inkomati Basin with focus on Improved Hydrological Understanding for Risk-based Operational Water Management. Deliverable 3, Project K5/1935, September 2011.

Conference proceedings

A.M.L. Saraiva Okello, S. Uhlenbrook, G. Jewitt, E. S. Riddell, I. Masih, P. van der Zaag (2014) Using tracers to develop a holistic understanding of runoff generation in a large semi-arid basin in Southern Africa, Paper presented at PhD Symposium, UNESCO-IHE, Delft, 29-30 September 2014.

A.M.L. Saraiva Okello, G. Jewitt, I. Masih, S. Uhlenbrook, P. van der Zaag (2013) Global change issues in the Incomati River Basin, Paper presented at Global Water Systems Project International Conference on Water in the Anthropocene, Bonn, Germany, 21-24 May 2013.

A.M.L. Saraiva Okello, E. Riddell, S. Uhlenbrook, I. Masih, G. Jewitt, P. van der Zaag, S. Lorentz. Isotopic and Hydrochemical River Profile of the Incomati River Basin. Conference proceedings 13th WaterNet/WARFSA/GWP-SA Symposium, Johanesburg, South Africa 2012.

E.S. Riddell, **A.M.L. Saraiva Okello**, P. Van der Zaag, G.P.W. Jewitt, S. Uhlenbrook, B. Jackson, T.K. Chetty. Risk-based operational water management through improved hydrological understanding to augment IWRM institutional capacity in the Incomati, Conference proceedings 12th WaterNet/WARFSA/GWP-SA Symposium, Maputo, Mozambique, 2011

Saraiva A. M. L., Wenninger J. Uhlenbrook S. and Ndlovu L. Experimental investigation of water fluxes in irrigated sugarcane in Swaziland using environmental isotopes, Conference proceedings 11th WaterNet/WARFSA/GWP-SA Symposium, Victoria Falls, Zimbabwe, 2010

Saraiva A. M. L., Tilmant A. Uhlenbrook S., Van der Zaag P. Risk-based operational water management for the Incomati River Basin, poster presented at 11th WaterNet/WARFSA/GWP-SA Symposium, Victoria Falls, Zimbabwe, 2010

Netherlands Research School for the
Socio-Economic and Natural Sciences of the Environment

D I P L O M A

For specialised PhD training

The Netherlands Research School for the
Socio-Economic and Natural Sciences of the Environment
(SENSE) declares that

Aline Maraci Lopes Saraiva Okello

born on 10 March 1983 in Maputo, Mozambique

has successfully fulfilled all requirements of the
Educational Programme of SENSE.

Delft, 2 May 2019

The Chairman of the SENSE board	the SENSE Director of Education
Prof. dr. Martin Wassen	Dr. Ad van Dommelen

The SENSE Research School has been accredited by the Royal Netherlands Academy of Arts and Sciences (KNAW)

K O N I N K L I J K E N E D E R L A N D S E
A K A D E M I E V A N W E T E N S C H A P P E N

The SENSE Research School declares that Aline Maraci Lopes Saraiva Okello has successfully fulfilled all requirements of the Educational PhD Programme of SENSE with a work load of 36.5 EC, including the following activities:

SENSE PhD Courses

- Environmental research in context (2012)
- Research in context activity: 'Co-organizing PhD Symposium at IHE Delft (1-5 October 2012) and acting as member of the IHE PhD Association Board, PAB (October 2012-September 2014)"

Other PhD and Advanced MSc Courses

- Python and land surface modelling with PCRaster, IHE Delft (2013)

Selection of External workshops and training

- Remote Sensing workshop, University of KwaZulu Natal, South Africa (2011)
- Coursera MOOC - Water Supply and sanitation policy in developing countries, University of Manchester, United Kingdom (2014)
- Coursera MOOC - Sustainability in practice, University of Pennsylvania, United States in America (2014)

Management and Didactic Skills Training

- Organization and reporting of the Incomati Basin Science Symposium special session, 4-7 November 2012, Drakensberg, South Africa
- RISKOMAN project closing workshop organization, 25 November 2014, Maguga, Swaziland
- Supervising MSc student with thesis entitled 'Identification and Quantification of Runoff Components in the Kaap Catchment, South Africa' (2014)

Selection of Oral Presentations

- *Isotopic and Hydrochemical River Profile of the Incomati River Basin.* 13th Waternet/WARFSA/GWP-SA Symposium, 31 October- 2 November 2012, Johannesburg South Africa
- *Global change issues in the Incomati River Basin.* Global Water Systems Project International Conference on Water in the Anthropocene, 21-24 May 2013, Bonn, Germany
- *Using tracers to develop a holistic understanding of runoff generation in a large semi-arid basin in Southern Africa*. Boussineq Lecture, 23 October 2014, Amsterdam, The Netherlands

SENSE Coordinator PhD Education

Dr. Peter Vermeulen